McGraw-Hill

Mathematics

Daily Homework Practice

McGraw-Hill School Division
A Division of The ***McGraw-Hill*** *Companies*

McGraw-Hill School Division
Two Penn Plaza
New York, New York 10121-2298

Printed in the United States of America

ISBN 0-02-100287-8 / 5

4 5 6 7 8 9 024 05 04 03 02 01

GRADE 5
Contents

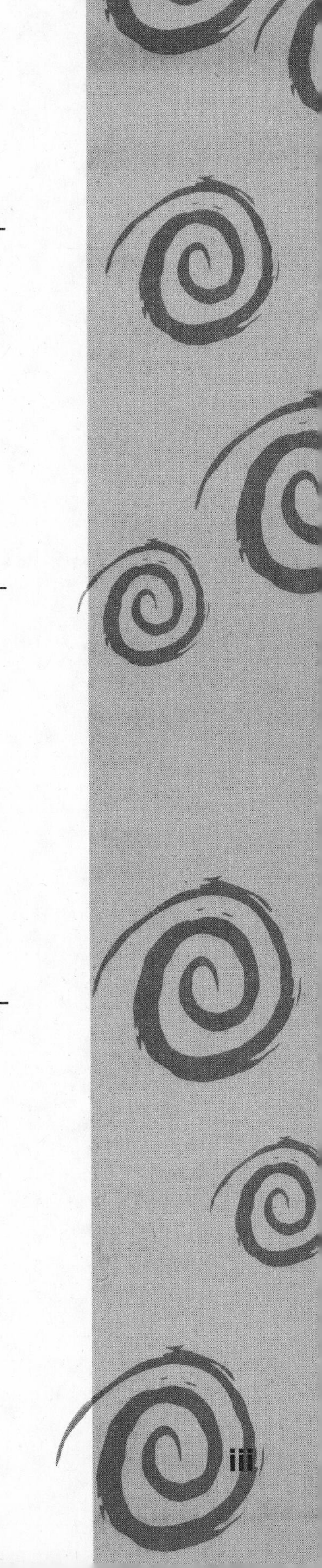

Chapter 1: Place Value: Add and Subtract Whole Numbers and Decimals

Chapter 2: Multiply Whole Numbers and Decimals

Chapter 3: Divide Whole Numbers and Decimals

Chapter 4: Data, Statistics, and Graphs

Chapter 5: Number Theory and Fraction Concepts

Chapter 6: Add and Subtract Fractions

Chapter 7: Multiply and Divide Fractions

Chapter 8: Measurement

Chapter 9: Integers

Chapter 10: Algebra: Expressions and Equations

Chapter 11: Geometry

Chapter 12: Perimeter, Area, and Volume

Chapter 13: Ratio and Probability

Chapter 14: Percents

Name ______________________________

Place Value Through Billions

Name the place and value of each underlined digit.

1. 1,2<u>4</u>9

____

____

2. 1<u>4</u>,191

____

____

3. <u>3</u>,475, 222,892

____

____

4. 10,4<u>2</u>6,865

____

____

5. 8<u>8</u>,776,554

____

____

6. 30,1<u>2</u>3,548,004

____

____

Write each number in standard form.

7. 23 million, 864 thousand, 200 ______________

8. 212 billion, 804 million, 34 thousand, 401 ______________

Write the numbers in expanded form.

9. 42,800 ______________________________

10. 2,354,476 ______________________________

Problem Solving

11. China's population is 1,246,871,951. How do you read that number?

____

____

12. China's total land area is 3,705,392 square miles. Russia's total land area is 6,592,745 square miles. How much more land does Russia have than China?

____

Source: *The New York Times Almanac 2000*

Spiral Review

Write each number in standard form.

13. 12 million, 785 thousand, 324 ______________

14. 29 billion, 487 million, 123 thousand, 973 ______________

Name ____

1·2 Explore Decimal Place Value

Write the decimal and the fraction.

1. 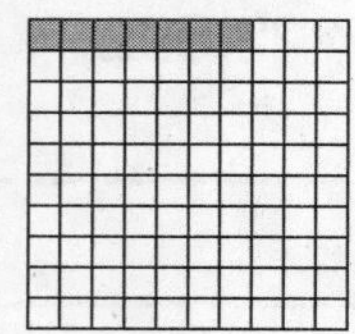

____

2. 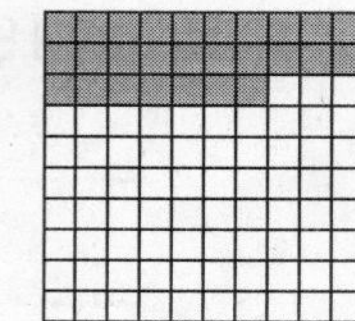

____

3. 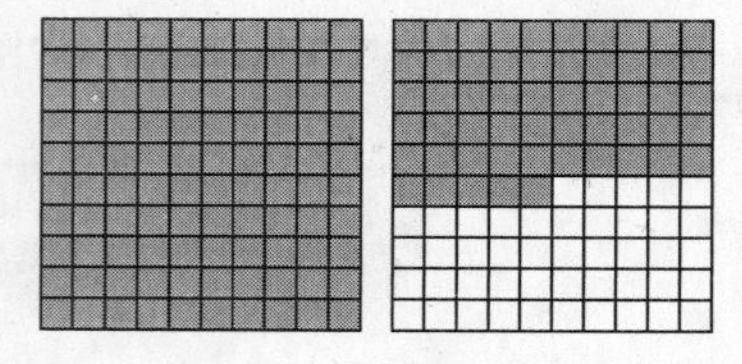

____

4. 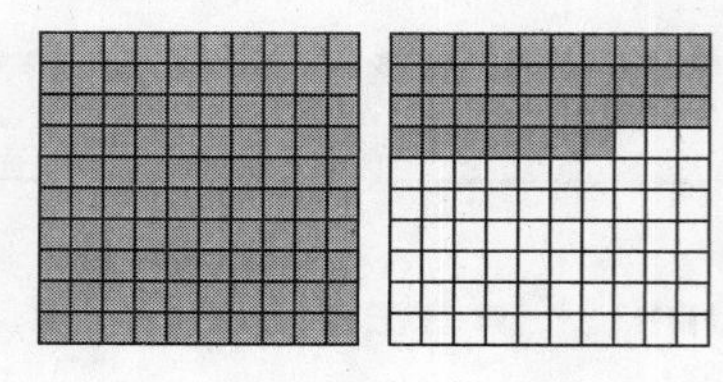

____

Solve.

5. Lola and Mark are planting a vegetable garden. They want to plant lettuce in 0.25 of the garden. Use the grid below to model how much of the garden will be lettuce. Then write the decimal as a fraction.

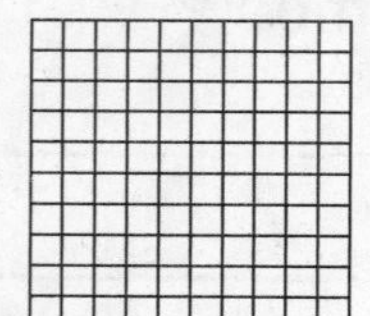

____

6. Sue has a coin collection. She has 100 coins in all and 56 of them are nickels. Use the grid to model the number of nickels Sue has. Then write a decimal and a fraction to represent the number of nickels she has.

____

Spiral Review

Name the place and value of each underlined digit.

7. 43,0<u>4</u>4

____

8. 4<u>4</u>,096,781

____

9. 10,348,00<u>7</u>

____

10. 38,75<u>1</u>,429

____

Name ______________________________

1•3 Decimal Place Value

Name the place of the underlined digit.

1. 1.2̲3 ____________
2. 3.14̲6 ____________
3. 8.654̲ ____________

4. 0.0163̲ ____________
5. 2,7̲26.78 ____________
6. 47.513̲ ____________

Write in standard form.

7. seven and 8 tenths ____________

8. fourteen and 75 hundredths ____________

Write the word name and expanded form.

9. 4.70 ______________________________

10. 34.045 ______________________________

Name an equivalent decimal.

11. 0.8 ____________
12. 2.14 ____________
13. 210.06 ____________

Problem Solving

14. In 1992, Quincy Watts of the United States won the 400-meter track event with a time of forty-three and five tenths seconds. Write this number.

____

Source: *The World Almanac and Book of Facts 2000*

15. In 1994 Al Unser, Jr., reached a speed of 160.872 miles per hour in the Indianapolis 500 automobile race. Write an equivalent decimal for this number.

____

Spiral Review

Name the place and the value of each underlined digit.

16. 3,6̲73 ____________ ____________
17. 78̲,133 ____________ ____________
18. 88̲,114,567 ____________ ____________

Name ______________________________

1·4 Compare and Order Whole Numbers and Decimals

Compare. Write >, <, or =.

1. 1,722 ________ 1,723

2. 12,444,832 ________ 9,457,998

3. 212,783,806 ________ 212,793,806

4. 1.7 ________ 1.67

5. 4.560 ________ 4.506

6. 35.800 ________ 35.8

Order from least to greatest.

7. 413,540; 9,912; 84,127

____

8. 218,521; 218,512; 218,512.7

____

9. 6.347; 6.437; 6.4; 6.34; 6

____

10. 0.67; 0.671; 0.608; 0.655; 0.635

____

Find two numbers between the numbers given.

11. 23.3 and 23.8 ______________________________

12. 6,841 and 6,846 ______________________________

Problem Solving

13. Philips Arena in Atlanta seats 20,000. The Charlotte Coliseum seats 24,042. The United Center in Chicago seats 21,711. Which is the largest stadium?

Source: *The World Almanac and Book of Facts 2000*

____

14. Frankie runs on his school's track team. The top 3 records set by his teammates are 15.02 seconds, 15.09 seconds, and 15.11 seconds. What is the slowest time Frankie can have to be in first place?

____

Spiral Review

Name the place of the underlined digit.

15. $17.6\underline{8}0$

____

16. $8.01\underline{7}7$

____

17. $18.\underline{1}14$

____

Name ____

Problem Solving: Reading for Math

Use the Four-Step Process

Solve. Use the four-step process.

The Verrazano-Narrows Bridge in New York is 4,260 feet long. The Williamsburg Bridge across the East River is 1,600 feet long and the Brooklyn Bridge, across the same river, is 1,595 feet long. Other suspension bridges include the Oakland Bay Bridge in San Francisco, California, 2,310 feet long; the Golden Gate Bridge in San Francisco, 4,200 feet long; and the Seaway Skyway in Ogdensburg, New York, which is 2,150 feet long.

1. Which bridge is the shortest? ____________________

2. List the three longest bridges in order from greatest to least.

____

Use data from the table for problems 3–4.

Underwater Vehicular Tunnels

Tunnel Name (Year Built)	Location	Waterway	Length (ft)
Ted Williams (1995)	Boston, MA	Boston Harbor	8,448
Queens Midtown (1940)	New York, NY	East River	6,414
Brooklyn-Battery (1950)	New York, NY	East River	9,117
Holland (1927)	New York, NY	Hudson River	8,557
Baltimore Harbor (1957)	Baltimore, MD	Baltimore Harbor	7,392

Source: *The World Almanac and Book of Facts 2000*

3. Which tunnel was built the earliest? ____________________

4. List the three longest tunnels in order from greatest to least.

____

Spiral Review

Compare. Write <, >, or =.

5. 28,234,688 ____ 28,204,669

6. 11,404,687 ____ 11,405,687

Name ______________________________

Add and Subtract Whole Numbers and Decimals

Add or subtract.

1. 21,125 − 11,465
2. 32,444 + 29,088
3. 148,809 + 58,911
4. 5,074 − 1,837
5. 56,396 − 39,989

6. 71.475 + 52.364
7. 388.56 − 97.68
8. 92.06 + 19.45
9. 42.706 − 0.844
10. $573.38 + 48.98

11. 614,782,596 + 368,425,491
12. 466.724 − 73.968
13. 48,763.612 − 9,483.713

Choose the sum that is greater.

14. A. 3,680 + 308
 B. 3,670 + 385

15. A. 470,000 + 180,000
 B. 490,000 + 220,000

16. A. 1.346 + 0.318
 B. 1.664 + 0.98

17. A. 41.650 + 8.48
 B. 44.580 + 2.561

Problem Solving

18. In 1996, the United States women's 400-meter relay team beat its own 1992 record by 0.16 second. The 1992 time was 42.11 seconds. What was its record time in 1996?

 ____

Source: *The Time Almanac 2000*

19. If adult tickets to an Olympic event cost $25.50 and a child's ticket costs $20.75, how much will a family of 2 adults and 2 children pay for tickets to the event?

 ____

Spiral Review

Compare. Write >, <, or =.

20. 21.47 ____ 21.5
21. 27.17 ____ 2.710
22. 4.086 ____ 4.10
23. 75.600 ____ 75.6
24. 59.218 ____ 59.208
25. 689.2 ____ 6,892

Name ______________________________

Estimate Sums and Differences

Round to the underlined place.

1. $\underline{2}$,864 ________
2. 57,7$\underline{8}$7 ________
3. 84.$\underline{5}$3 ________
4. 36,8$\underline{0}$7 ________
5. 54.6$\underline{9}$4 ________

6. 34.$\underline{7}$5 ________
7. 74,57$\underline{7}$,048 ________
8. 2.00$\underline{3}$6 ________
9. 3,$\underline{8}$97 ________
10. 1$\underline{4}$9.98 ________

Round to the place indicated.

11. 848 (hundreds) ________
12. 5,742 (thousands) ________
13. 1,875 (tens) ________

Estimate each sum or difference. Show your work.

14. $\begin{array}{r} 5.37 \\ +\ 0.08 \\ \hline \end{array}$

15. $\begin{array}{r} 0.86 \\ +\ 3.077 \\ \hline \end{array}$

16. $\begin{array}{r} 45.29 \\ -\ 10.742 \\ \hline \end{array}$

Problem Solving

17. A well-known television actor makes $44,775 per episode on a weekly television show. About how much does he earn for a 13-week series?

____

18. In 1999, there were 356 different species of animals on the endangered list and 578 species of plants. Estimate these numbers to the nearest ten. Then find the total number of endangered species to the nearest hundred.

____

____

Source: *The New York Times Almanac 2000*

Spiral Review

Write two equivalent decimals for each.

19. 5.8 ________
20. 0.7 ________
21. 8.05 ________

Name ______________________________

Problem Solving: Strategy

Find a Pattern

Find a pattern to solve.

1. A high school student begins practicing the high jump with the bar set at 3 feet 5 inches and raises the bar 0.5 inch after each successful jump. How high will the bar be after 4 successful jumps? ______________

2. A high school pole vaulter raises the bar for her vaults each time she makes a successful jump. If the bar is set at 9 feet 6 inches and she raises the bar 1.5 inches after each jump, how high will the bar be after 6 successful jumps? ______________

3. Velma jogs 1 lap around the track each day for two weeks. In the third and fourth weeks, she jogs 3 laps each day around the track. During the next two weeks, she runs 6 laps each day. During the fifth and sixth weeks, her goal is to jog 10 laps each day. If this pattern continues, how many laps will she run each day during the seventh and eighth weeks? ______________

Mixed Strategy Review

4. In the 1996 Olympic 200-meter run, a United States runner won with a 19.32 seconds. In the 1992 200-meter run, the record time was 20.01 seconds. By how much did the record improve from 1992 to 1996?

   ____

5. The longest women's throws for the javelin in the Olympics are 68.40 meters, 74.66 meters, and 69.56 meters. Write these lengths in order from least to greatest.

   ____

Source for both: *The New York Times Almanac 2000*

Spiral Review

Compare. Write <, >, =.

6. 21,455,876 ______ 21,456,900

7. 234,859,102 ______ 234,857,404

Name ____________________

Properties of Addition

Identify the addition property used to rewrite each problem.

1. 8 + 47 + 23 = 47 + 8 + 23

____

2. (3 + 1.73) + 0.37 = 3 + (1.73 + 0.37)

____

3. 32 + (18 + 28) = (32 + 18) + 28

____

4. 336 + 884 = 884 + 336

____

5. 472 + 0 = 472

____

6. 47 + 82 + 75 = 82 + 47 + 75

____

7. 25 + 87 + 54 = 25 + 54 + 87

____

8. 2 + 7 + 8 + 3 = (2 + 8) + (7 + 3)

____

Add or subtract. Describe your work.

9. 54 + 22 = ______

10. 683 + 87 = ______

11. 296 + 104 = ______

12. 529 – 371 = ______

13. 304 – 289 = ______

14. $512 – $36 = ______

15. 423 + 78 = ______

16. 276 – 193 = ______

17. 338 + 248 = ______

Problem Solving

18. The loudness of a whisper is measured at 20 decibels. The sound of a jet plane is measured at 140 decibels. How much louder is the plane than the whisper?

____

19. A turtle moves at a speed of about 0.17 mile per hour. A sloth moves at about 0.15 mile per hour, while a snail has a speed of only 0.03 mile per hour. How much faster is a turtle than a snail?

____

Spiral Review

Order from least to greatest.

20. 0.27, 0.43, 0.05, 1.15

____

21. 0.125, 0.625, 0.375

____

22. 93,909; 87,100; 78,997

____

Name ______________________________

2-1 Patterns of Multiplication

Copy and complete.

1. $2 \times 3 = 6$
$2 \times 30 = 60$
$2 \times 300 = n$
$2 \times 3{,}000 = n$

____

2. $3 \times 5 = 15$
$3 \times 50 = x$
$3 \times 500 = 1{,}500$
$3 \times 5{,}000 = x$

____

3. $6 \times 30 = 180$
$60 \times 30 = s$
$600 \times 30 = 18{,}000$
$6{,}000 \times 300 = s$

____

4. $1 \times \$40 = \$40$
$10 \times \$40 = \$400$
$100 \times \$40 = m$
$1{,}000 \times \$40 = m$

____

5. $3 \times 7 = d$
$30 \times 7 = 210$
$300 \times 7 = d$
$3{,}000 \times 7 = d$

____

6. $4 \times 6 = p$
$4 \times 60 = 240$
$4 \times 600 = p$
$4 \times 6{,}000 = p$

____

Multiply.

7. $8 \times 40 =$ ________ **8.** $5 \times 1{,}000 =$ ________ **9.** $200 \times 9 =$ ________

10. $10 \times 60 =$ ________ **11.** $120 \times 20 =$ ________ **12.** $60 \times 300 =$ ________

13. $800 \times 7 =$ ________ **14.** $700 \times 40 =$ ________ **15.** $2{,}600 \times 100 =$ ________

Problem Solving

16. Marcus is reading a book about Spain. He usually reads an average of 15 pages a day. He has 7 days to finish his book. How many pages can he read in that time? He has 100 pages left. Will he be able to finish the book in that time?

____

17. Spain has a land area of about 195,000 square miles. If there are about 200 people per square mile, how many people are there in Spain?

____

Spiral Review

Identify the addition property used to rewrite each problem.

18. $9 + 73 + 13 = 9 + 13 + 73$ ______________________________

19. $(2.1 + 1.7) + 3.6 = 2.1 + (1.7 + 3.6)$ ______________________________

Name ____________________

Explore the Distributive Property

Multiply.

1. $8 \times 38 =$ ______

2. $7 \times 28 =$ ______

3. $9 \times 31 =$ ______

4. $5 \times 39 =$ ______

5. $4 \times 98 =$ ______

6. $7 \times 297 =$ ______

7. $32 \times 8 =$ ______

8. $99 \times 9 =$ ______

9. $796 \times 6 =$ ______

Rewrite each problem using the Distributive Property.

10. 5×58 ______

11. 8×29 ______

12. 7×19 ______

13. 2×37 ______

14. 9×28 ______

15. 4×49 ______

Solve.

16. Junius has 4 tulip beds. Each bed has 48 tulip plants. How many tulips in all does Junius have in his garden?

______

17. Nora is on a committee for a school party. There will be 4 party favors for each guest. If 198 guests have been invited, how many favors should Nora order?

______

Spiral Review

Round each number to the underlined place.

18. 56.66<u>4</u>8 ______

19. 6.<u>7</u>87 ______

20. 428.<u>0</u>62 ______

Name ______________________________

2·3 Multiply Whole Numbers

Multiply.

1. $475 \times 3 =$ ______
2. $62 \times 43 =$ ______
3. $137 \times 46 =$ ______
4. $59 \times 8 =$ ______
5. $86 \times 82 =$ ______
6. $1{,}206 \times 23 =$ ______
7. $3{,}465 \times 18 =$ ______
8. $704 \times 7 =$ ______
9. $654 \times 68 =$ ______

10. 245×6
11. 333×16
12. 36×48
13. 498×92
14. $3{,}542 \times 48$
15. $10{,}772 \times 8$
16. $44{,}333 \times 5$
17. $2{,}168 \times 88$

Compare. Write >, <, or =.

18. 6×48 __ 9×56
19. 8×82 __ 6×96
20. 65×174 __ 46×208
21. 28×345 __ 32×298
22. 92×28 __ 184×14
23. 399×46 __ 466×50

Problem Solving

24. If it costs $1,155 to fly from New York City to Uganda round trip, how much would it cost for 6 round-trip tickets? ______________

25. Mt. Kilimanjaro in Tanzania is 19,340 feet high. Mt. Margherita in Uganda is 16,750 feet high. How much higher is Mt. Kilimanjaro than Mt. Margherita?

____

Spiral Review

26. $19.82 + 9.7 =$ ______
27. $27.176 - 18.72 =$ ______
28. $31.2 - 8.8 =$ ______
29. $181.56 + 28.1 =$ ______

Name ____________________

Properties of Multiplication

Multiply. Name the property you used.

1. $5 \times 37 =$ ______

______

______

2. $1 \times 97 =$ ______

______

______

3. $6 \times 55 \times 5 =$ ______

______

______

4. $7 \times 297 =$ ______

______

______

5. $21{,}888 \times 0 =$ ______

______

______

6. $12{,}005 \times 4 =$ ______

______

______

Problem Solving

7. The population of New Zealand is 3,662,265. What is that number rounded to the nearest hundred thousand? ______

8. The population of Auckland, New Zealand, is 997,940. The population of Wellington is 335,468. How much greater is the population of Auckland than that of Wellington? ______

Spiral Review

Compare. Write >, <, or =.

9. 2.1 ___ 2.01

10. 0.430 ___ 0.4300

11. 0.807 ___ 0.087

12. 3.15 ___ 3.1

Name ________________________________

Estimate Products of Whole Numbers and Decimals

Estimate by rounding.

1. 9.2×10 ________
2. 11×63 ________
3. 132×43 ________
4. 212×82 ________
5. 61×39 ________
6. $2{,}029 \times 42$ ________
7. 8.7×11.1 ________
8. 129.4×40 ________
9. 8.22×490 ________
10. 1.78×27
11. 874×9
12. 368×488
13. 8.3×78

Estimate by clustering.

14. $387 + 372 + 416$ ________
15. $82.8 + 79.2 + 77$ ________
16. $1{,}100 + 1{,}205 + 1{,}212$ ________
17. $897 + 912 + 889$ ________
18. $1{,}490 + 1{,}398 + 1{,}421$ ________
19. $3.79 + 4.16 + 4.21$ ________

Estimate by rounding. Write > or <.

20. 37.92×1.8 ___ 78.76
21. 58.89×21.07 ___ $1{,}600$
22. $340 ___ $80.45 × 3.7
23. 78.6×4.2 ___ $92 + 89 + 88$

Problem Solving

24. Milagros bought a toy for $14.99, two postcards for $2.15 each, and a poster for $8.99. She gave the clerk $30. Estimate her change, if there is no sales tax. __________

25. The population of Greenland was estimated to be fify-nine thousand, eight hundred twenty-seven. Write that number in expanded form.

____

Spiral Review

Find the number that makes each sentence true.

26. $72 + 8 + 3 = (72 + \_\_\_\_) + 8$
27. $376 + \_\_\_\_ = 376$
28. $6 \times (43 + 7) = (6 \times 43) + (\_\_\_\_ \times 7)$
29. $184 + \_\_\_\_ = 754 + 184$

Name ____________________

Problem Solving: Reading for Math
Estimate or Exact Answer

Solve. State whether the problem requires an estimate or an exact answer.

1. Marina plans to visit Uruguay next summer. She knows that there are approximately 49 people per square mile in that country. If the land area is 68,000 square miles, how many people live in Uruguay?

   ____

2. About 91 out of every 100 people live in the cities of Uruguay. Express that number as a decimal and as a fraction.

   ____

3. Students in the United States paid a customs fee of $108 to send boxes of English-language books to Uruguay. If the cost was $4.50 per box of books, how many boxes did they send?

   ____

4. The population of Asunción, the capital of Paraguay, is estimated to be 546,600. The population of Montevideo, the capital of Uruguay, is about 1,303,200. How many more people live in Montevideo than live in Asunción?

   ____

5. The border shared by Paraguay and Brazil is 1,290 kilometers in length. If a traveler were to drive 250 kilometers each day, about how many days would it take to start from one end and end at the other?

   ____

6. Tourism in Uruguay brings in about eight hundred ninety-five million dollars each year. Write that number in standard form.

   ____

Spiral Review

Estimate by rounding.

7. 9.8×19.9 ________ **8.** 41.8×3.12 ________ **9.** 62.1×7.23 ________

Name ______________________________

Multiply Whole Numbers by Decimals

Multiply.

1. 8.4 × 5   **2.** 14.4 × 7   **3.** 6.73 × 8   **4.** 39.86 × 9   **5.** 6.04 × 100

6. 0.6 × 6 = ________   **7.** $41.86 × 5 = ________   **8.** 4.9 × 11 = ________

9. 42.86 × 7 = ________   **10.** 342.6 × 4 = ________   **11.** 12.85 × 10 = ________

12. 74.85 × 5 = ________   **13.** 61.05 × 3 = ________   **14.** 0.43 × 8 = ________

Find the multiple of 10 that makes each statement true.

15. 3.4 × ________ = 340   **16.** ________ × 2.5 = 25

17. 0.58 × ________ = 580   **18.** 26.4 × ________ = 2,640

Compare. Write <, >, or =.

19. 3.6 × 2 ___ 0.8 × 10   **20.** 0.7 × 100 ___ 3.5 × 20   **21.** 10.1 × 4 ___ 8.1 × 5

Problem Solving

22. Suppose that 800 tourists buy gifts at a museum gift shop each day. If each tourist spends an average of $15.50, how much will the gift shop take in per day? ____________

23. Naomi buys 2 maps of Jordan at $19.95 each and a guide book for $25.50. How much does she spend in all? ____________

Spiral Review

24. 487.65 + 128.24 + 43.17   **25.** 64.87 − 8.82   **26.** 83.66 + 9.77   **27.** 823.45 − 110.33

Name ______________________________

Explore Multiplying Decimals by Decimals

Multiply.

1. 0.6 × 0.4 = ______ **2.** 0.3 × 0.9 = ______ **3.** 0.3 × 0.3 = ______

4. $\begin{array}{r} 0.4 \\ \times\, 0.5 \\ \hline \end{array}$ **5.** $\begin{array}{r} 0.8 \\ \times\, 0.6 \\ \hline \end{array}$ **6.** $\begin{array}{r} 0.7 \\ \times\, 0.5 \\ \hline \end{array}$ **7.** $\begin{array}{r} 0.1 \\ \times\, 0.6 \\ \hline \end{array}$ **8.** $\begin{array}{r} 0.5 \\ \times\, 0.6 \\ \hline \end{array}$

9. 0.7 × 0.1 = ______ **10.** 0.4 × 0.9 = ______ **11.** 0.4 × 0.8 = ______

12. 0.7 × 0.3 = ______ **13.** 0.6 × 0.6 = ______ **14.** 0.9 × 0.9 = ______

15. 0.2 × 0.1 = ______ **16.** 0.3 × 0.4 = ______ **17.** 0.8 × 0.2 = ______

18. 0.4 × 0.7 = ______ **19.** 0.6 × 0.7 = ______ **20.** 0.1 × 0.1 = ______

21. 0.5 × 0.5 = ______ **22.** 0.6 × 0.3 = ______ **23.** 0.4 × 0.4 = ______

Solve.

24. One half, or 0.5, of a box of 24 oranges were eaten. How many oranges were left in the box? ______

25. One fourth, or 0.25, of the trees in an orchard were flowering. If there were 224 trees in the orchard, how many were in flower? ______

Spiral Review

26. 7.5 × 6 = ______ **27.** 8.2 × 8 = ______

28. 4.7 × 3 = ______ **29.** 4 × 1.8 = ______

Name ____________________

Multiply Decimals by Decimals

Multiply.

1. 3.42×0.5

2. 72.84×6.2

3. 10.11×0.7

4. 2.22×3.3

5. 4.48×5.73

6. 204.62×0.4

7. 6.07×8.4

8. 23.1×1.58

9. 3.14×2.27

10. 1.12×1.01

11. $3.9 \times 0.4 =$ ________ **12.** $4.4 \times 0.7 =$ ________ **13.** $23.5 \times 0.5 =$ ________

14. $1.5 \times 0.4 =$ ________ **15.** $14.88 \times 2.8 =$ ________ **16.** $3.08 \times 1.4 =$ ________

Compare. Write <, >, or =.

17. 0.86×0.258 ___ 0.300

18. 1.5×0.6 ___ 1.8×0.5

19. 0.024×7 ___ $0.141 + 0.18$

20. 7.5×4 ___ 3.42×8

Problem Solving

21. The number of goats in Nepal is about twice the number of buffalo. If there are about 3,400,000 buffalo, how many goats are there?

____

22. In Nepal the unit of currency is the rupee. One U.S. dollar is equal to 68.23 rupees. How many rupees would you need to buy three dollars worth of cloth?

____

Spiral Review

23. $44.07 + 2.3 =$ ________

24. $2.4 - 1.58 =$ ________

25. $54.91 - 1.58 =$ ________

26. $43.82 + 3.87 =$ ________

Name ______________________________

Problem Solving: Strategy
Guess and Check

Use the guess and check strategy to solve.

1. Franco sees 8 camels and counts 14 humps in all. If all Bactrian camels have two humps and all Dromedary camels have one hump, how many of each type of camel did Franco see?

____

2. Ronna saw a photograph of a number of Bactrian and Dromedary camels near an oasis. She counted 27 humps in all. How many of each kind of camel were in the photograph?

____

3. Trisha is on a vacation. She sends letters and postcards to her friends at home. She needs 33¢ in postage for each letter and 20¢ for each postcard. She spends $3.58 in postage. How many letters and postcards did she send?

____

Mixed Strategy Review

4. Sam and Gina set up tables for a history club lunch discussion. A square table seats 4 people, one on each side. If they push the tables together end to end, how many tables will they need to seat 16 club members?

____

5. For a public meeting, the history club spent $37.00 on refreshments, $6.00 for film rentals, and $7.00 for programs. In all, they spent $140 for the meeting. Their only other expense was for rental of the meeting space. How much did they spend on rental?

____

Spiral Review

Name the place and value of the underlined number.

6. 385,<u>9</u>97

____

7. <u>2</u>,345,007

____

Name ____________________

Exponents

Rewrite using a base and exponent.

1. $3 \times 3 \times 3 =$ ______

2. $2 \times 2 \times 2 \times 2 \times 2 =$ ______

3. $4 \times 4 \times 4 \times 4 =$ ______

4. $8 \times 8 \times 8 \times 8 \times 8 =$ ______

5. $6 \times 6 \times 6 \times 6 \times 6 \times 6 \times 6 =$ ______

6. $9 \times 9 \times 9 \times 9 \times 9 \times 9 \times 9 \times 9 =$ ______

7. $7 \times 7 \times 7 =$ ______

8. $5 \times 5 \times 5 \times 5 \times 5 \times 5 =$ ______

Write in standard form.

9. 7^2 ______ **10.** 3^4 ______ **11.** 4^6 ______ **12.** 2^7 ______ **13.** 9^0 ______

14. 3^5 ______ **15.** 11^4 ______ **16.** 10^6 ______ **17.** 18^1 ______ **18.** 8^5 ______

19. 1^0 ______ **20.** 14^3 ______ **21.** $(0.2)^4$ ______ **22.** $(0.01)^2$ ______ **23.** $(0.4)^3$ ______

Problem Solving

24. If a phone team asks each person to call two people, how many rounds will it take to call 64 people? ______________

25. Members of a tour group take a bus at 8:00 A.M. They travel for 2 hours 35 minutes.

They spend 2 hours 45 minutes at their tour site and then return by the same route. What time will they get back to their hotel? ______________

26. A type of bacteria reproduces by splitting in two every thirty minutes. How many bacteria will there be after 4 hours if there are 2 bacteria to begin with?

____

Spiral Review

27. $4.17 \times 8.1 =$ ______ **28.** $1.7 \times 0.92 =$ ______ **29.** $4.2 \times 6.2 =$ ______

Name ________________________________

3-1 Relate Multiplication and Division

Complete the fact family.

1. $4 \times 6 = 24$
   $6 \times 4 = 24$
   ______ $\div 6 = 4$
   ______ $\div 4 = 6$

2. $3 \times 9 = 27$
   $9 \times 3 = 27$
   $27 \div 3 =$ ______
   $27 \div 9 =$ ______

3. $5 \times 6 = 30$
   ______ $\times 5 = 30$
   $30 \div$ ______ $= 5$
   $30 \div 5 = 6$

4. $9 \times 8 = 72$
   $8 \times 9 = 72$
   $72 \div$ ______ $= 8$
   $72 \div$ ______ $= 9$

5. $8 \times 4 = 32$
   $4 \times 8 = 32$
   ______ $\div 4 = 8$
   ______ $\div 8 = 4$

6. $4 \times 12 = 48$
   $12 \times 4 = 48$
   $48 \div$ ______ $= 4$
   ______ $\div 4 = 12$

Divide.

7. $36 \div 9 =$ ____
8. $49 \div 7 =$ ____
9. $28 \div 4 =$ ____
10. $72 \div 8 =$ ____
11. $21 \div 7 =$ ____
12. $16 \div 4 =$ ____
13. $20 \div 4 =$ ____
14. $56 \div 7 =$ ____

15. $6\overline{)48}$
16. $12\overline{)36}$
17. $9\overline{)54}$
18. $6\overline{)42}$
19. $2\overline{)16}$
20. $2\overline{)14}$
21. $9\overline{)81}$
22. $12\overline{)84}$

Problem Solving

23. The duck-billed platypus has a bill about 9 centimeters long. Its body measures about 36 centimeters in length, and its tail is about 18 centimeters long. How many times longer than its bill is the body of the platypus?

____

24. How many times longer is its body than its tail? ______________________

Spiral Review

Rewrite using a base and an exponent.

25. $9 \times 9 \times 9$ ______
26. $2 \times 2 \times 2 \times 2$ ______
27. 8×8 ______

Name ____________________

Explore Dividing by 1-Digit Divisors

Divide.

1. $4\overline{)26}$ **2.** $9\overline{)48}$ **3.** $7\overline{)32}$ **4.** $8\overline{)57}$

5. $3\overline{)29}$ **6.** $5\overline{)108}$ **7.** $8\overline{)98}$ **8.** $6\overline{)124}$

9. $7\overline{)64}$ **10.** $6\overline{)158}$ **11.** $7\overline{)87}$ **12.** $8\overline{)196}$

13. $6\overline{)50}$ **14.** $4\overline{)39}$ **15.** $5\overline{)56}$ **16.** $4\overline{)47}$

17. $6\overline{)125}$ **18.** $7\overline{)93}$ **19.** $3\overline{)16}$ **20.** $4\overline{)147}$

21. $9\overline{)83}$ **22.** $4\overline{)87}$ **23.** $5\overline{)44}$ **24.** $7\overline{)120}$

Solve.

25. Wanda spent $34.72 on groceries and $18.75 on gasoline. She had $70.00 when she left home. How much money does she have left after her purchases? ____________

26. Orange juice costs $3.25 a half gallon. How much will 3 half gallons cost? ____________

Spiral Review

27. $(0.04) \times (0.2) =$ ________ **28.** $0.26 \times 0.37 =$ ________

29. $0.83 \times 1.25 =$ ________ **30.** $0.44 \times 4.4 =$ ________

31. $6.4 \times 1.8 =$ ________ **32.** $5.5 \times 2.4 =$ ________

Name ______________________________

Divide by 1-Digit Divisors

Divide.

1. $4\overline{)324}$ **2.** $6\overline{)410}$ **3.** $7\overline{)256}$ **4.** $8\overline{)384}$

5. $3\overline{)843}$ **6.** $4\overline{)850}$ **7.** $8\overline{)437}$ **8.** $6\overline{)1{,}743}$

9. $8\overline{)2{,}117}$ **10.** $7\overline{)2{,}349}$ **11.** $5\overline{)2{,}443}$ **12.** $3\overline{)1{,}196}$

13. $2\overline{)1{,}847}$ **14.** $4\overline{)3{,}870}$ **15.** $5\overline{)1{,}234}$ **16.** $4\overline{)5{,}072}$

17. 43,064 ÷ 6 = ____________ **18.** 13,893 ÷ 6 = ____________

Problem Solving

19. A marsupial mole has a mass of about 60 grams. The cus cus has a mass about 60 times as great. What is the mass of the cus cus in grams? ____________

20. A wombat weighs 37,448 grams, a kangaroo weighs 63,582 grams, and a Tasmanian devil weighs 10,854 grams. Estimate their total weight, to the nearest hundred grams. ____________

Spiral Review

Estimate the sum.

21. 37.4 + 73.6 ____________ **22.** 1.68 + 0.76 ____________

23. 2,842 + 18 ____________ **24.** 1.187 + 4.753 ____________

Name ________________________________

Divide by 2-Digit and 3-Digit Divisors

Divide. Check your answer.

1. $32\overline{)4,748}$ **2.** $65\overline{)7,512}$ **3.** $27\overline{)8,248}$ **4.** $342\overline{)2,130}$

5. $76\overline{)2,498}$ **6.** $54\overline{)6,564}$ **7.** $82\overline{)9,545}$ **8.** $62\overline{)48,136}$

9. $32,665 \div 62 =$ ____________ **10.** $23,784 \div 44 =$ ____________

Find each dividend.

11. ____________ $\div 42 = 39$ R5 **12.** ____________ $\div 28 = 362$ R16

Decide whether the first digit of the quotient is too high or too low. Then complete.

13. $36\overline{)1,304}$ with 4 above ____________

14. $76\overline{)3,214}$ with 3 above ____________

15. $132\overline{)54,400}$ with 5 above ____________

Problem Solving

16. The whale shark may grow to be 18 meters long. The shortest type of shark may be only 10 centimeters long. How many times longer is the whale shark? Hint: There are 100 centimeters in a meter. ____________

17. It takes Marcia 15 minutes to sew a patch on the sleeve of a club shirt. If she spent 2 hours and 45 minutes sewing patches, how many patches did she sew on? ____________

Spiral Review

18. $842.3 + 32.7 =$ ________ **19.** $0.636 - 0.337 =$ ________

20. $1.78 + 0.48 + 9.22 =$ ________ **21.** $3.266 - 0.008 =$ ________

Name ______________________________

3•5 Estimate Quotients

Complete the pattern.

1. $18 \div 3 = 6$
   $180 \div 3 =$ ______
   $1,800 \div 3 =$ ______
   $18,000 \div 3 =$ ______

2. $56 \div 7 = 8$
   ______ $\div 7 = 80$
   $5,600 \div$ ______ $= 800$
   $56,000 \div 7 =$ ______
   $560,000 \div 7 =$ ______

3. $24 \div 4 = 6$
   $240 \div 40 =$ ______
   $2,400 \div$ ______ $= 60$
   $24,000 \div 40 =$ ______
   $240,000 \div 40 =$ ______

4. $72 \div 12 = 6$
   $720 \div$ ______ $= 60$
   $7,200 \div 120 =$ ______
   $72,000 \div 120 =$ ______
   $720,000 \div$ ______ $= 6,000$

Divide.

5. $350 \div 7 =$ ______
6. $42,000 \div 60 =$ ______
7. $180 \div 6 =$ ______
8. $70,000 \div 10 =$ ______
9. $200 \div 4 =$ ______
10. $30,000 \div 600 =$ ______

Estimate. Use compatible numbers.

11. $642 \div 9$ ______
12. $552 \div 7$ ______
13. $3,574 \div 9$ ______

Problem Solving

14. Yvonne is reading a book about the animals of Africa. She has 123 pages left to read. If she reads about 20 pages each day, how many days will it take her to finish the book? ______

15. Madagascar has an area of about 226,700 square miles. If there are about 66 people per square mile, estimate the total population.

____

Source: *The Time Almanac 2000*

Spiral Review

Compare. Use >, <, or =.

16. 7.23 ______ 7.19
17. 0.04879 ______ 0.4879
18. 3.46 ______ 3.64

Name ____________________

Problem Solving: Reading for Math

Interpreting the Remainder

Solve. Tell how you interpreted the remainder.

1. On a field trip, 185 students and teachers take buses to a zoological park. If each bus holds 42 persons, how many buses are needed for the trip?

   ____

2. At the park, there is a special exhibit of exotic insects. No more than 35 persons can go through the exhibit at one time. How many exhibit periods will it take for all the students and teachers to go through the exhibit?

   ____

3. A special lecture-film show takes 35 minutes. If the park is open for 8 hours each day, how many complete showings will there be during that time? Hint: 8 hours = 480 minutes.

   ____

4. The cafeteria at the park has tables that seat 10 people. How many tables will it take to seat these185 persons?

   ____

5. The students took 58 photographs and plan to display them for their classmates at school. If they can put 10 photographs on each poster, how many posters will they need?

   ____

6. At the science fair, the students plan to give short talks about their experience at the zoological garden. They can seat only 42 people for their talk. How many times must the students give their talk if they expect 246 students and parents to attend?

   ____

Spiral Review

7. $4\overline{)245}$ **8.** $5\overline{)787}$ **9.** $8\overline{)116}$ **10.** $3\overline{)646}$

Name ______________________________

Problem Solving: Strategy
Work Backward

Solve. Use the work-backward strategy.

1. A bird club counts birds in the woods around their town. They see 20 baby birds and count 26 adult female birds. They know that there are an equal number of adult male and adult female birds present. How many birds did the students count in all?

____

2. The student council wants to raise money to plant flowers around the school. They decide to sell raffle tickets. They raise $320 in all. The first 20 tickets were sold at $3 each. The rest of the tickets cost $2. How many $2 tickets did they sell?

____

3. During March, the price of rose bushes tripled. In April, the price dropped $0.75 per bush. By May, the price was cut in half. In June, the price rose $0.25 to $1.75 per bush. What was the price at the beginning of March?

____

Mixed Strategy Review

4. Mark and Letty play a number game. Mark tells Letty to pick a number, add 19 to it, then multiply the sum by 2, and then subtract 7 from the product. Letty gets 125. What number did she start with? __________

5. A group of 26 people want to sit together at a long table. Each table seats 3 people on a long side and one person on a short side. If they place the short sides end to end, how many tables will it take to seat the group at one long table? __________

Spiral Review

6. $45 \times 200 =$ ______ **7.** $50 \times 4{,}000 =$ ______ **8.** $80 \times 500 =$ ______

9. $70 \times 300 =$ ______ **10.** $25 \times 2{,}000 =$ ______ **11.** $60 \times 600 =$ ______

Name ____________________

3-8 Divide Decimals by Whole Numbers

Divide. Round each quotient to the nearest hundredth if necessary.

1. $7\overline{)9.8}$ **2.** $3\overline{)3.9}$ **3.** $5\overline{)6.5}$ **4.** $8\overline{)65.28}$

5. $5\overline{)3.8}$ **6.** $10\overline{)4.2}$ **7.** $6\overline{)9.72}$ **8.** $3\overline{)24.42}$

9. $4\overline{)1.86}$ **10.** $100\overline{)482}$ **11.** $6\overline{)288}$ **12.** $5\overline{)78}$

13. 75.7 ÷ 2 = ______ **14.** 9.25 ÷ 5 = ______ **15.** 22.19 ÷ 7 = ______

16. 76 ÷ 6 = ______ **17.** 135 ÷ 4 = ______ **18.** 114 ÷ 10 = ______

Complete the pattern.

19. 72.9 ÷ 10 = ______
72.9 ÷ 100 = 0.729
72.9 ÷ 1,000 = ______

20. 5.8 ÷ 10 = ______
5.8 ÷ 100 = ______
5.8 ÷ 1,000 = ______

21. 45 ÷ 10 = ______
45 ÷ 100 = ______
45 ÷ 1,000 ______

22. 25 ÷ 10 = ______
25 ÷ 100 = ______
25 ÷ 1,000 = ______

23. 46 ÷ 10 = ______
46 ÷ 100 = ______
46 ÷ 1,000 = ______

24. 9.8 ÷ 10 = ______
9.8 ÷ 100 = ______
9.8 ÷ 1,000 = ______

Problem Solving

25. Lincoln learns from his science textbook that all insects have 6 legs. How many legs would 8,000 insects have? ______

26. A millipede having 100 segments is 19.8 centimeters long. How long would each equal segment be? Round to the nearest tenth of a centimeter. ______

Spiral Review

Express using a base and an exponent.

27. 4 × 4 × 4 × 4 ______ **28.** 5 × 5 × 5 ______ **29.** 8 × 8 × 8 × 8 × 8 ______ **30.** 9 × 9 ______

Name ________________________________

Explore Dividing Decimals by Decimals

Divide.

1. $5.6 \div 0.8 =$ ______ **2.** $8.4 \div 1.2 =$ ______ **3.** $4.5 \div 0.9 =$ ______

4. $6.4 \div 0.8 =$ ______ **5.** $5.2 \div 1.3 =$ ______ **6.** $1.6 \div 0.4 =$ ______

7. $9.1 \div 0.7 =$ ______ **8.** $2.4 \div 0.4 =$ ______ **9.** $1.4 \div 0.2 =$ ______

10. $1.8 \div 0.3 =$ ______ **11.** $3.6 \div 0.9 =$ ______ **12.** $6.5 \div 1.3 =$ ______

13. $7.5 \div 0.5 =$ ______ **14.** $8.1 \div 2.7 =$ ______ **15.** $2.1 \div 0.3 =$ ______

16. $5.4 \div 0.6 =$ ______ **17.** $7.6 \div 0.4 =$ ______ **18.** $10.5 \div 1.5 =$ ______

19. $14.4 \div 1.2 =$ ______ **20.** $6.6 \div 1.1 =$ ______ **21.** $0.64 \div 0.08 =$ ______

Solve.

22. Sara has 30 yards of fabric. She and her friends want to make matching skirts. If it takes 1.5 yards for each skirt, how many skirts can they make from Sara's fabric? ______________

23. Alicia has 5 pennies, 5 quarters, and 5 dimes. Marcus has 18 pennies, 4 quarters, and 6 dimes. Selena has 9 pennies, 8 quarters, and 9 dimes. Order the three students by the amount of money each has, from least to greatest.

____

Spiral Review

24. $75 + 139 =$ ______ **25.** $308 - 49 =$ ______

26. $900 \times 80 =$ ______ **27.** $4 \times 5 \times 18 \times 5 =$ ______

28. $36 + 148 =$ ______ **29.** $548 - 64 =$ ______

30. $600 \times 50 =$ ______ **31.** $6 \times 4 \times 3 \times 5 =$ ______

Name ____________________

Divide Decimals by Decimals until there is no remainder.

Divide.

1. 1.8 ÷ 0.4 = ______ **2.** 1.47 ÷ 0.7 = ______ **3.** 8.16 ÷ 2.4 = ______

Divide. Round to the nearest tenth.

4. $1.6\overline{)4.28}$ **5.** $4.2\overline{)76.25}$ **6.** $3.8\overline{)82.5}$ **7.** $2.6\overline{)12.64}$

8. $1.6\overline{)58}$ **9.** $0.32\overline{)36.36}$ **10.** $5.5\overline{)485.88}$ **11.** $6.2\overline{)8{,}624.6}$

12. $14.62\overline{)273.54}$ **13.** $1.86\overline{)66.24}$ **14.** $6.08\overline{)77.216}$ **15.** $5.12\overline{)60.42}$

Divide. Round to the nearest hundredth.

16. $5.8\overline{)20.76}$ **17.** $5.5\overline{)74.85}$ **18.** $10.8\overline{)315.4}$ **19.** $6.6\overline{)123.7}$

20. $2.44\overline{)182.8}$ **21.** $0.48\overline{)68}$ **22.** $0.95\overline{)15.55}$ **23.** $1.7\overline{)675}$

Problem Solving

24. If a spotted turtle needs about 1.3 acres of land in order to survive, how many turtles could there be in an area that covers 31.2 acres? ______

25. A rectangle has an area of 534.75 square inches. If its width is 15.5 inches, what is its length? ______

Spiral Review

26. 4.7 + 11.4 = ______ **27.** 4.04 × 4.04 = ______ **28.** 12.6 − 4.56 = ______

29. 28.6 + 4.2 = ______ **30.** 3.06 × 8.82 = ______ **31.** 534.8 − 3.9 = ______

Name 

Explore Collecting, Organizing, and Displaying Data

Use the data in this table to answer the questions that follow.

Candace's Survey		
Telephones at Home	**Fifth Graders Surveyed**	
	Tally	**Frequency**
1	卌	5
2	卌 ||	7
3	卌 |||	
4		3
5	卌	5
6 or more	||	

1. Fill in the missing data on Candace's table.

2. Make a line plot of Candace's data.

Number of Telephones at Home

3. Describe the data in the line plot in one sentence.

____

4. How many fifth-graders were surveyed? ________

Spiral Review

5. $34 \times 47 =$ ________ **6.** $308 \times 67 =$ ________ **7.** $74 \times 407 =$ ________

8. $56 \times 9{,}002 =$ ________ **9.** $30 \times 4{,}200 =$ ________ **10.** $26 \times 512 =$ ________

Name ________________________________

Range, Mode, Median, and Mean

Find the range, mode, median, and mean.

1. 1, 1, 2, 4, 5, 1, 2, 1, 2, 6, 8

____

2. 1.3, 1.4, 1.5, 1.5, 1.3, 1.4, 1.4, 1.4

____

3. 500, 550, 455, 475, 5000, 555, 700, 610, 524

____

4. Why is the median so much smaller than the mean in Problem 3?

____

Problem Solving

5. Alexa had the following scores on her math quizzes: 57, 84, 81, 97, 81, 89. She can choose among the mean, the median, or the mode for her quiz grade. Which one should she choose?

____

Use data in the table for problems 6–7. Round to the nearest tenth.

Ages of Workers in a Fast Food Restaurant

17	18	17	18	17	19	17	18	18	23	24	25	49

6. What is the mean age of all the workers? ________________

7. The manager hires a new worker who is 47 years old. What is the new mean age of the entire work crew? ________________

Spiral Review

8. $2.3 \times 45 =$ ________ **9.** $4.3 \times 5.6 =$ ________ **10.** $1.05 \times 24.3 =$ ________

Name ________________________________

4-3 Read and Make Pictographs

Use the pictograph below to answer questions 1–4.

Ice Cream Cones Sold Today

Vanilla	
Chocolate Chip	
Cookie Dough	
Butter Crunch	
Rocky Road	

Each [cone] represents 10 ice cream cones

1. Which flavor was most popular?

   ____

2. How many cones of that flavor were sold? __________

3. Rocky Road outsold Butter Crunch by how many cones? __________

4. How many ice cream cones were sold all together? __________

5. Survey your house for books. Count how many books are in each room of your home. Make a frequency table to display the data. Make a pictograph for the data.

   ____

6. The mean number of pages in the books in Janie's bedroom is 252 pages. The mean number of book pages in the kitchen is 1,221 pages, and the mean number of book pages in the living room is 342 pages. What is the mean number of book pages in these three rooms?

   ____

Spiral Review

Round each number to nearest tenth.

7. 1.34 ________ **8.** 1.07 ________ **9.** 2.045 ________

Round each number to nearest hundredth.

10. 1.367 ________ **11.** 2.785 ________ **12.** 3.8924 ________

Name ____

Read and Make Bar Graphs

Use data from the graph for problems 1–5.

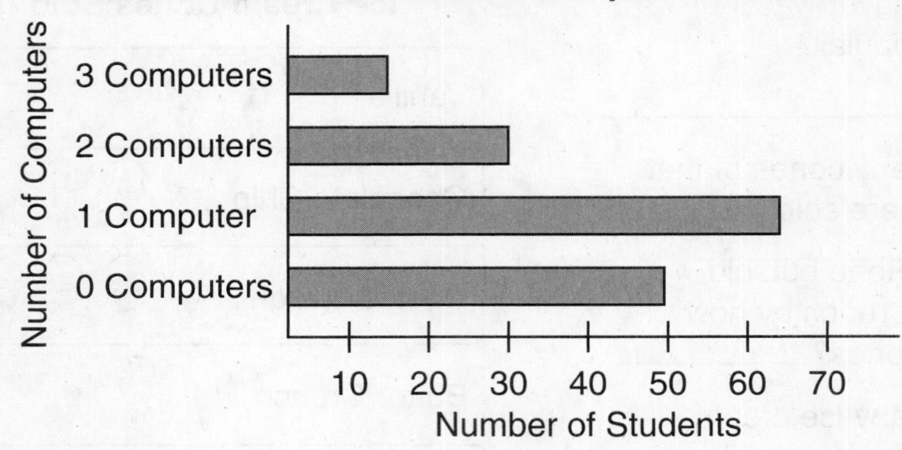

1. How many students have no computer at home? ________
2. How many students have one computer at home? ________
3. What is the range of the data? ________
4. How many students were surveyed? ________
5. How many computers are in the homes of all the students surveyed? ________

Problem Solving

Use data from the table for problems 6–7.

6. Find the range, mode, median, and mean for these temperatures.

   ____

7. Which two cities are closest in mean January temperatures?

   ____

Mean January Temperatures (°F)

City	Temperature
Honolulu	80
Boston	36
Chicago	29
Miami	75
Houston	62

Spiral Review

8. $12 \div 0.3 =$ ________
9. $1.5\overline{)4.5510}$
10. $\frac{1.6}{0.8} =$ ________

Name ____________________

Read and Make Histograms

Use data from the histogram for problems 1–3.

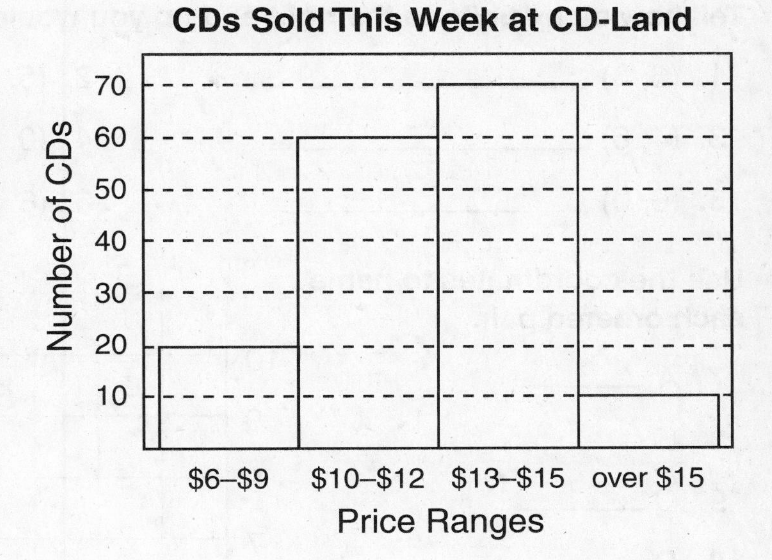

1. What was the most common price range for CDs sold this week? ____________

2. What was the least common price range for CDs sold this week? ____________

3. How many CDs were sold this week at CD Land?

____

Problem Solving

4. Make a histogram for data in the table.

Ages of people at the park

Age	Number
0–14 years	18
15–29 years	24
30–44 years	6
45–60 years	10

5. Why are the bars in a histogram touching and not separated?

____

Spiral Review

Order from least to greatest

6. 0.345, 3.456, 0.0345, 34.5, 0.3451 ____________

7. 1.01, 1.015, 1.035, 10.35, 0.13, 0.1 ____________

Name ____

Read and Make Line Graphs

Tell how many units to the right and up you would move for each coordinate pair.

1. (6, 3) ____________________ **2.** (5, 2) ____________________

3. (4, 6) ____________________ **4.** (0, 3) ____________________

5. (6, 0) ____________________ **6.** (8, 8) ____________________

Use the coordinates to name each ordered pair.

7. *A* ________

8. *B* ________

9. *C* ________

10. *D* ________

11. *E* ________

12. *F* ________

13. *G* ________

14. *H* ________

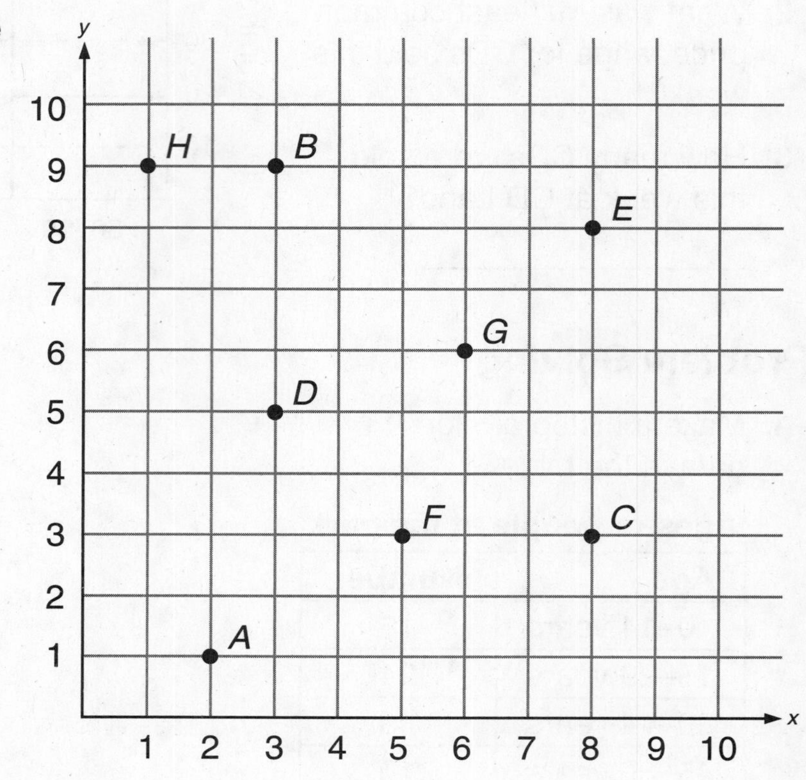

Problem Solving

15. An airplane leaves an airport at (2,7). It lands 6 units east of that point. What are its new coordinates? ________

16. A pilot leaves an airport at (4, 2) and files to (4, 9). In what direction is she flying? ________

Spiral Review

17. 3(6 + 2) = ________ **18.** 4(7 + ________) = 28

19. 5(________ + 4) = 40 **20.** ________(6 + 2) = 64

Name ____________________

Problem Solving: Reading for Math
Change Scales

Use data from the graph for problems 1–4.

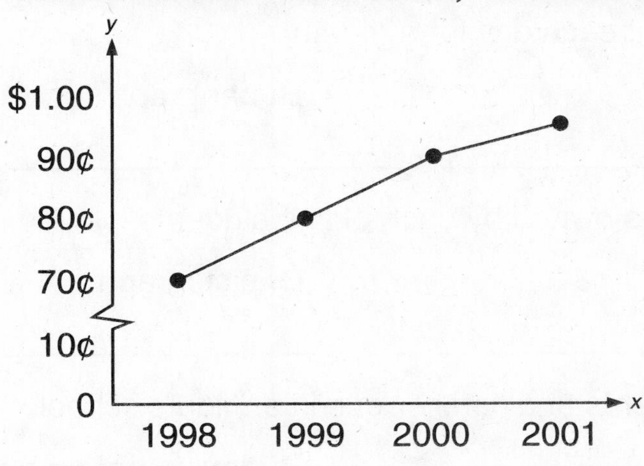

1. What does the graph seem to show about the price of a bus ticket?

____

2. On a separate sheet of paper, redraw the graph without the break in the vertical axis. What does the new graph seem to show about the price of a bus ticket?

____

3. On another sheet of paper, redraw the graph again, keeping the vertical axis interval distance the same but using 5¢ intervals instead of 10¢ intervals. How does this new graph compare with the one you just drew?

____

4. Suppose you wanted to show that the price of a bus ticket hardly changed during 1998–2001. How would you change the vertical axis interval distance of the original graph? ____________________

Spiral Review

5. 0.36 + 23 = ________
6. 2.368 + 4.08 = ________
7. 873.7 + 19.48 = ________
8. 0.43 + 5 + 4.6 = ________

Name ____

Problem Solving: Strategy

Make a Graph

For problems 1–4, choose the graph that best displays the data. Explain your choice.

1. Hours of sunlight each day for a month

A. line graph **B.** bar graph

____

2. Number of shoes owned by each of five students

A. line graph **B.** pictograph

____

3. Number of students of different ages in a middle school

A. histogram **B.** double bar graph

____

Mixed Strategy Review

Use data from the table for problems 4–5.

Water Depth at Different Times of Day in Boston Harbor	
Time	Water Depth (feet)
12:00	10
1:00	12
2:00	13.5
3:00	15
4:00	17
5:00	20
6:00	21

4. What is the best type of graph to display the data in the table? Explain.

____

____

5. Does water depth increase at an equal rate?

____

____

Spiral Review

6. 8.432 – 2.4 = ________ **7.** 12.62 – 4.354 = ________

8. 8 – 2.352 = ________ **9.** 17.863 – 16.9 = ________

Name ____________________

Read and Make Stem-and-Leaf Plots

Use data from this stem-and-leaf plot for problems 1–6.

Number of Minutes Spent in a Grocery Store by Shoppers

stems	leaves
3	4 4 6 6 9 9
2	5 5 5 6 8 8 8
1	1 5 7 7 9
0	5 8 9

1. What is the range of time spent in the store? ____________
2. What is the least amount of time spent in the store? ____________
3. How many shoppers are represented in the stem-and-leaf plot? ____________
4. What is the total number of shopping minutes represented? ____________
5. What was the mean of time spent shopping? ____________
6. What was the median time spent shopping? ____________

Problem Solving

7. On a separate sheet of paper, make a stem-and-leaf plot of the following data.

____

8. Alice watered her garden twice on every sunny day during the year shown. How many times did she water her garden?

____

Sunny Days in a Year	
January	8
February	12
March	18
April	16
May	22
June	26
July	27
August	23
September	18
October	21
November	10
December	12

Spiral Review

Give the range, median, mean, and mode for the following.

9. 12, 15, 13, 15, 11, 10, 8

____

10. 7, 4, 6, 12, 4, 4, 6, 7, 8

____

Name ____________________

Sampling

Name each population and sample.

1. Survey ten players on the soccer team about their favorite shoes.

____

2. Survey 100 students out of a school with 633 students.

____

3. Survey 20 people in the school band about their favorite music.

____

A researcher wants to find out what type of backpack is the most popular at a school. Tell whether or not each sample is a random sample. Is it a representative sample? Explain.

4. The researcher talks to every tenth student on a list of all the students in the school. ____________________

____

5. The researcher talks to the students who signed up to be interviewed.

____

____

Problem Solving

6. Jim surveys 64 people. Jill surveys 112 people. How many times more people did Jill survey?

____

7. A survey reveals that 35 students rode their bikes to school in March. In May, 54 students rode their bikes to school. How many more students rode their bikes in May than in March?

____

Spiral Review

8. $4 - 2.135 =$ ________
9. $6.2 \times 7.004 =$ ________
10. $1.3 + 0.794 + 12.005 =$ ________
11. $12{,}000 \div 4.8 =$ ________

Name ______________________________

5-1 Divisibility

Of 2, 3, 5, 6, 9, and 10, list which numbers each of these numbers is divisible by.

1. 48 ______________

2. 72 ______________

3. 90 ______________

4. 180 ______________

5. 200 ______________

6. 67 ______________

7. 12,531 ______________

8. 6,939 ______________

9. 144 ______________

10. 1,946 ______________

11. 4,635 ______________

12. 9,123 ______________

13. 2,001 ______________

14. 546 ______________

15. 440 ______________

16. 150 ______________

17. 720 ______________

18. 684 ______________

19. 714 ______________

20. 406 ______________

Problem Solving

21. The graduating class at Blaine High School has 328 members. They can sit in rows of 6, 8, or 12 at graduation. Which row size should they choose so that no student is left over? ________

22. Jennifer has baked 24 cookies. She plans to sell them in bags of 2, 3, or 5 cookies. What size bags can she use and not have any cookies left over? ________

Spiral Review

23. $8 - 2.37 =$ ________

24. $5.6 + 7.91 =$ ________

25. $12.9 - 4.65 =$ ________

26. $18.23 + 7.9 =$ ________

27. $3.4 - 0.63 =$ ________

28. $5.4 + 8.62 =$ ________

29. $29.7 - 3.62 =$ ________

30. $94.37 + 4.8 =$ ________

Name ______________________________

Explore Primes and Composites

Write a prime factorization for each number. Use exponents if you can. Tell if each number is prime or composite.

1. 20 ______________ **2.** 23 ______________

3. 38 ______________ **4.** 91 ______________

5. 72 ______________ **6.** 125 ______________

7. 40 ______________ **8.** 96 ______________

9. 53 ______________ **10.** 54 ______________

11. Use a factor tree to show the prime factorization of 60. ______________

12. Use a factor tree to show the prime factorization of 80. ______________

Solve.

13. Jim is thinking of a number. He says that if he multiplies his number by 2, multiplies it by 2 again, and then multiplies it by 2 one more time, his answer will be 24. What number is he thinking of?

____

14. Frank wants to plant 24 tomato plants in rows of equal numbers of plants. What are three different ways he can plant his tomatoes in equal rows?

____

____

Spiral Review

15. 4(3 + 5) = ________ **16.** 5(6 + ________) = 45

17. 4(________ + 3) = 28 **18.** ________ (4 + 5) = 72

19. 6(5 + ________) = 48 **20.** 3(7 + 2) = ________

Name ______________________________

Common Factors and Greatest Common Factor

Find the GCF of the numbers.

1. 12 and 36 ______

2. 16 and 20 ______

3. 33 and 55 ______

4. 40 and 90 ______

5. 18 and 36 ______

6. 24 and 54 ______

7. 17 and 23 ______

8. 18 and 44 ______

9. 54 and 81 ______

10. 42 and 28 ______

11. 56 and 16 ______

12. 13 and 39 ______

13. 30 and 24 ______

14. 35 and 56 ______

15. 70 and 55 ______

16. 24 and 56 ______

17. 18 and 42 ______

18. 15 and 50 ______

19. 8 and 32 ______

20. 28 and 49 ______

Problem Solving

21. Tonya and Heather picked flowers in Tonya's garden to sell at the local grocery store. Tonya picked 64 flowers and Heather picked 40. They want to sell the flowers in bunches with an equal number of flowers. What is the greatest number of flowers they can put in a bunch? ______________

22. A store sells pencils in packs of six or eight. Mr. Garcia, who teaches one class of 30 students and another class of 24 students, wants to buy pencils for all of his students. What size packs should he buy so that he will have no pencils left over? ______________

Spiral Review

Find the mean for each set of numbers.

23. 26, 29, 24, 28, 33 ______

24. 40, 42, 39, 46, 50, 47 ______

25. 15, 13, 17, 15 ______

26. 60, 66, 63, 62, 64 ______

Name ______________________________

5·4 Fractions

Name each fraction shown.

1.

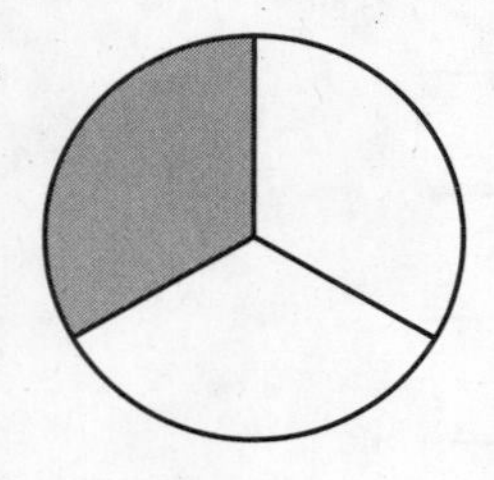

____

2.

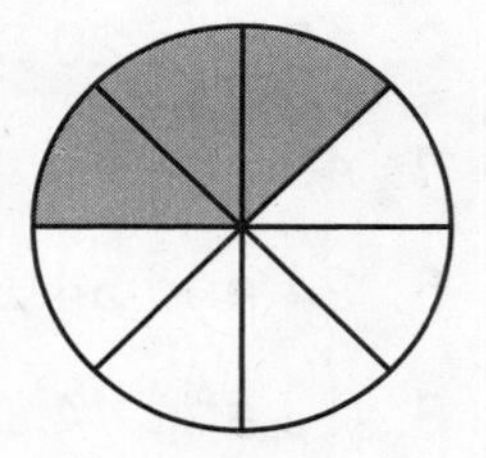

____

3.

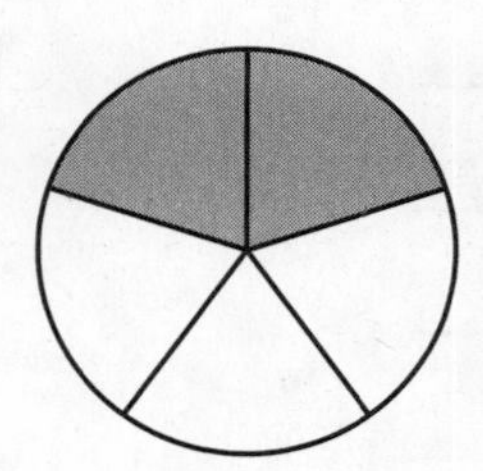

____

4.

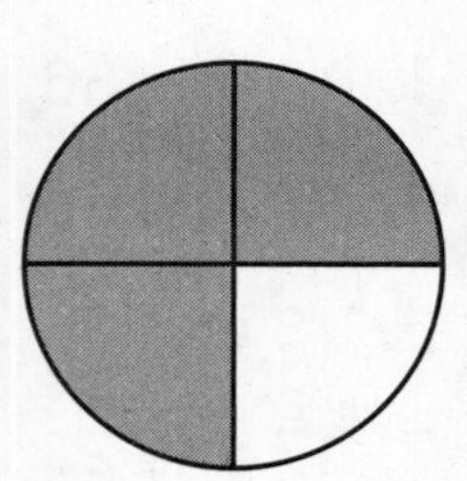

____

Draw a model to show each fraction.

5. $\frac{2}{3}$

6. $\frac{1}{8}$

7. $\frac{3}{5}$

Write two equivalent fractions for each fraction.

8. $\frac{3}{5}$ ______________________________

9. $\frac{5}{8}$ ______________________________

Problem Solving

10. In a class, four out of 20 students own a pet bird. What fraction of the class own birds? _______

11. Anne has a collection of 30 old coins. Two-fifths of the coins are quarters. How many quarters are there in Anne's coin collection? _______

Spiral Review

Find the median for each set.

12. 16, 12, 58, 8, 17, 47, 95 _______

13. 12, 18, 23, 18, 24 _______

14. 18, 52, 48, 86, 50 _______

15. 24, 30, 18, 22, 29, 27 _______

Name ____________________

Problem Solving: Reading for Math

Extra and Missing Information

Meadville Recreation Center Craft Courses

Courses	Sign-ups	FEES
Pottery	15	**Residents**
Wood Carving	12	$25 per course
Jewelry Making	18	Maximum $100
Flower Arranging	10	**Nonresidents**
Glassblowing	20	$30 per course
		Maximum $125

Solve. If there is not enough information, write *not enough information.*

1. What fraction of all the sign-ups are for glassblowing? ______
2. How many students are taking jewelry making? ______
3. How many different students have signed up for pottery and flower arranging? ______
4. Betty wants to take the pottery course. How much will that cost her? ______
5. Tina wants to take two courses and has $55. Does she have enough money to pay the fee? ______
6. Emily is a resident and wants to take three courses. How much will that cost her? ______

Spiral Review

Write the value of the underlined digit in each number.

7. $3.\underline{2}65$ ______

8. $\underline{7}48.43$ ______

9. $12.6\underline{8}6$ ______

10. $1\underline{8}5,326$ ______

Name ____________________

Simplify Fractions

Write each fraction in simplest form

1. $\frac{2}{6}$ ____ 2. $\frac{3}{9}$ ____ 3. $\frac{6}{12}$ ____

4. $\frac{10}{15}$ ____ 5. $\frac{16}{20}$ ____ 6. $\frac{6}{8}$ ____

7. $\frac{8}{40}$ ____ 8. $\frac{24}{30}$ ____ 9. $\frac{10}{60}$ ____

10. $\frac{12}{48}$ ____ 11. $\frac{24}{36}$ ____ 12. $\frac{30}{48}$ ____

13. $\frac{16}{36}$ ____ 14. $\frac{6}{30}$ ____ 15. $\frac{8}{48}$ ____

16. $\frac{6}{24}$ ____ 17. $\frac{18}{24}$ ____ 18. $\frac{30}{45}$ ____

Problem Solving

Use data from the table for problems 19–20.

19. What fraction of all the class pets are fish, in simplest terms? ____

20. What fraction of all the class pets are dogs, in simplest terms? ____

Class Pets

Pet	Number
dog	10
cat	8
bird	6
fish	4
hamster	2

Spiral Review

Of 2, 3, 5, 6, 9, and 10, list which numbers each of these numbers is divisible by.

21. 48 ____________ **22.** 56 ____________

23. 64 ____________ **24.** 67 ____________

25. 24 ____________ **26.** 30 ____________

Name ________________________________

Least Common Multiple and Least Common Denominator

Find the LCM of the numbers.

1. 3 and 6 ______
2. 3 and 4 ______
3. 7 and 3 ______
4. 9 and 5 ______
5. 6 and 15 ______
6. 6 and 16 ______
7. 4 and 5 ______
8. 6 and 8 ______
9. 6 and 9 ______
10. 3 and 10 ______
11. 4 and 7 ______
12. 5 and 15 ______

Find the LCD for each pair of fractions.

13. $\frac{1}{2}$ and $\frac{1}{5}$ ______
14. $\frac{2}{3}$ and $\frac{3}{4}$ ______
15. $\frac{5}{6}$ and $\frac{2}{9}$ ______
16. $\frac{3}{5}$ and $\frac{1}{3}$ ______
17. $\frac{2}{3}$ and $\frac{4}{5}$ ______
18. $\frac{1}{6}$ and $\frac{2}{3}$ ______

Write equivalent fractions using the LCD.

19. $\frac{2}{3}$ and $\frac{3}{4}$ ______
20. $\frac{2}{5}$ and $\frac{1}{3}$ ______

Problem Solving

21. Hot dogs are sold in packages of 10. Hot dog rolls are sold in packages of 8. What is the least number of hot dogs Stan can serve at a cookout and not have any hot dogs or hot dog rolls left over? ______

22. Keyshawn is planning his soccer practice and his piano practice for the next 30 days. He plans to practice his piano every third day and to practice soccer every other day. How many days will he practice soccer and piano on the same day? ______

Spiral Review

23. 3.56×8.6

24. 0.16×0.024

25. $12 \div .04 =$ ______

Name ____________________

5-8 Compare and Order Fractions

Compare. Write >, <, or =.

1. $\frac{1}{5}$ ______ $\frac{1}{3}$

2. $\frac{2}{3}$ ______ $\frac{3}{4}$

3. $\frac{7}{9}$ ______ $\frac{3}{5}$

4. $\frac{3}{15}$ ______ $\frac{1}{5}$

5. $\frac{7}{12}$ ______ $\frac{3}{9}$

6. $\frac{2}{3}$ ______ $\frac{8}{12}$

Order from least to greatest.

7. $\frac{1}{2}, \frac{1}{5}, \frac{1}{3}$ ______

8. $\frac{1}{8}, \frac{2}{3}, \frac{1}{4}$ ______

9. $\frac{2}{3}, \frac{2}{5}, \frac{2}{4}$ ______

10. $\frac{3}{8}, \frac{1}{3}, \frac{2}{4}$ ______

11. $\frac{2}{3}, \frac{5}{9}, \frac{7}{8}$ ______

12. $\frac{3}{4}, \frac{8}{12}, \frac{1}{2}$ ______

Problem Solving

13. In Ms. Cruz's class, $\frac{2}{3}$ of the students play only video games, $\frac{3}{5}$ play only board games, and $\frac{7}{8}$ play only outdoor games. Which group has the largest number of students? ____________________

14. Fran, Julio, and Doreen have entered a walkathon to raise money for a local community group. After two hours, Julio has walked $\frac{2}{3}$ of the course, Doreen has walked $\frac{7}{9}$ of the course, and Fran has walked $\frac{9}{10}$ of the course. Who is closest to the finish? ____________________

Spiral Review

15. 18×56.3

16. $1{,}046 \times 45$

17. $1{,}800 \div 24 =$ ______

Name ____________________

Relate Fractions and Decimals

Write each decimal as a fraction in simplest form.

1. 0.25 ______ **2.** 0.4 ______ **3.** 0.15 ______

4. 0.75 ______ **5.** 0.02 ______ **6.** 0.3 ______

Write each fraction as a decimal.

7. $\frac{1}{2}$ ______ **8.** $\frac{1}{4}$ ______ **9.** $\frac{1}{5}$ ______

10. $\frac{1}{8}$ ______ **11.** $\frac{2}{5}$ ______ **12.** $\frac{3}{4}$ ______

13. $\frac{4}{5}$ ______ **14.** $\frac{7}{8}$ ______ **15.** $\frac{5}{8}$ ______

Is each pair of numbers equivalent? Write *yes* or *no*.

16. $\frac{7}{20}$; 0.35 ______ **17.** $\frac{8}{25}$; 0.36 ______

18. $\frac{17}{50}$; 0.34 ______ **19.** $\frac{3}{4}$; 0.65 ______

Problem Solving

20. The plans for a picnic table call for screws that are each 0.75 inch long. In simplest form, what fraction of an inch is this? ______

21. Jamie's weight at his checkup was $63\frac{1}{4}$ pounds. What is the decimal form of his weight? ______

Spiral Review

22. $2.5 \times 6.3 =$ ______ **23.** $9 \div 1.5 =$ ______

24. $7.77 \div 7 =$ ______ **25.** $0.4 \times 1.5 =$ ______

Name ______________________________

Problem Solving: Strategy
Make a Table

Use the make-a-table strategy to solve.

A video store kept track of the number of video games it rented each week.

Number of Video Games Rented

Week	Number	Week	Number	Week	Number
1	39	6	48	11	38
2	47	7	52	12	49
3	52	8	54	13	43
4	53	9	60	14	38
5	37	10	52	15	36

1. In what fraction of the weeks was the number of games rented between 31 and 40? Write the fraction in simplest form. _____

2. In what fraction of the weeks was the number of games rented greater than 50? Write the fraction in simplest form. _____

3. The store manager has discovered a mistake in the records. Week 1 video game rentals were actually 41, not 39. Now, in what fraction of the weeks was the number of games rented between 31 and 40? Write the fraction in simplest form. _____

Mixed Strategy Review

4. By nearly closing time, a CD store's sales for the day were 100 CDs, of which 15 were jazz CDs. Then, one final customer bought 10 jazz CDs. What fraction of the day's total sales were jazz CDs? Write the fraction in simplest form. _____

5. During a one-hour television show, there were 12 minutes of commercials. What fraction of the hour was not commercials? Write the fraction in simplest form. _____

Spiral Review

Find the median.

6. 12, 18, 16, 22, 15 _______ **7.** 21, 3, 5, 8, 13, 24 _______

Name ________________________________

Mixed Numbers

Write each mixed number as an improper fraction.

1. $2\frac{2}{3}$ ______ **2.** $4\frac{3}{5}$ ______ **3.** $6\frac{5}{6}$ ______

4. $7\frac{1}{3}$ ______ **5.** $5\frac{4}{5}$ ______ **6.** $2\frac{3}{4}$ ______

Write each improper fraction as a mixed number, in simplest form.

7. $\frac{7}{2}$ ______ **8.** $\frac{16}{4}$ ______ **9.** $\frac{17}{3}$ ______

10. $\frac{18}{2}$ ______ **11.** $\frac{22}{8}$ ______ **12.** $\frac{21}{9}$ ______

Write each mixed number as a decimal.

13. $3\frac{1}{2}$ ______ **14.** $4\frac{3}{4}$ ______ **15.** $7\frac{1}{5}$ ______

Write each decimal as a mixed number in simplest form.

16. 4.25 ______ **17.** 6.4 ______ **18.** 7.75 ______

Problem Solving

19. The plans for a kite call for a piece of string 7.75 inches long. What is this measurement expressed as a mixed number in simplest form? ______

20. On a space mission, an astronaut spends 10 days and 6 hours in a space station. What is this time period in days, expressed as a mixed number in simplest form? ______

Spiral Review

21. 5(7 – 3) = ______ **22.** 4(7 – ______) = 20

23. 6(______ – 5) = 18 **24.** ______(8 – 4) = 32

Name ____________________

Compare and Order Fractions, Mixed Numbers, and Decimals

Compare. Write <, >, or =.

1. $1\frac{1}{4}$ ______ $1\frac{1}{2}$

2. $2\frac{3}{5}$ ______ $2\frac{1}{2}$

3. $4\frac{1}{2}$ ______ $4\frac{6}{10}$

4. $6\frac{1}{3}$ ______ $\frac{25}{4}$

5. $\frac{19}{4}$ ______ $4\frac{3}{8}$

6. $8\frac{2}{8}$ ______ $8\frac{1}{4}$

7. $2\frac{4}{5}$ ______ $2\frac{3}{4}$

8. $7\frac{2}{3}$ ______ $7\frac{3}{4}$

Order from greatest to least.

9. $5\frac{2}{3}$, 5.9, $5\frac{3}{4}$ ____________

10. $7\frac{1}{3}$, 7.25, $7\frac{1}{5}$ ____________

11. $9\frac{2}{5}$, $9\frac{3}{4}$, $9\frac{1}{2}$ ____________

12. $4\frac{1}{8}$, $4\frac{1}{6}$, 4.2 ____________

Problem Solving

13. A certain brand of soap claims to make you "98 and $\frac{64}{100}$ percent clean." What is this mixed number in simplest terms? ________

14. In Seattle in September, it rained on 24 out of 30 days. In November, it rained on 0.6 of the total days in the month. Which month was rainier, September or November? ____________

Spiral Review

Round to the nearest tenth.

15. 3.25 ______

16. 6.143 ______

17. 8.27 ______

18. 1.976 ______

19. 7.44 ______

20. 5.362 ______

Name____

Add and Subtract Fractions and Mixed Numbers with Like Denominators

Add or subtract. Write your answer in simplest form.

1. $\frac{1}{5} + \frac{2}{5} =$ ______

2. $3\frac{4}{6} - 2\frac{1}{6} =$ ______

3. $5\frac{1}{4} + 2\frac{1}{4} =$ ______

4. $5\frac{2}{7} - 2\frac{1}{7} =$ ______

5. $4\frac{4}{9} + 2\frac{2}{9} =$ ______

6. $8\frac{7}{12} - 4\frac{3}{12} =$ ______

7. $7\frac{7}{8} - 4\frac{3}{8}$

8. $8\frac{5}{9} + 4\frac{1}{9}$

9. $6\frac{5}{6} - 2\frac{1}{6}$

10. $15\frac{7}{10} - 9\frac{3}{10}$

Problem Solving

11. $16\frac{3}{8} + 12\frac{5}{8}$

12. $18\frac{9}{12} - 14\frac{1}{12}$

13. $8\frac{5}{8} - 3\frac{1}{8}$

14. $29\frac{7}{12} - 22\frac{1}{12}$

15. Jolene needs to cut a piece of ribbon $4\frac{1}{4}$ in. long from a longer piece that is $10\frac{3}{4}$ in. long. How much ribbon will be left after she cuts off the piece she needs? ______

16. Carl was $52\frac{3}{8}$ in. tall at the start of the school year and $54\frac{7}{8}$ in. tall at the end of the school year. How many inches did Carl grow during the school year? ______

Spiral Review

Compare. Write <, >, or =.

17. $\frac{1}{10}$ ______ 0.1

18. $\frac{2}{9}$ ______ $\frac{3}{8}$

19. 4.25 ______ $4\frac{4}{5}$

Name ____________________

Problem Solving: Reading for Math

Choose an Operation

Solve. Tell how you chose the operation.

1. In the first week of January, rainfall totaled $1\frac{1}{16}$ in. In the second week, it totaled $1\frac{7}{16}$ in. How much more rain fell in the second week?

____

2. In the first two weeks of March, rainfall totaled $\frac{5}{8}$ in. In the third week, rainfall totaled $2\frac{1}{4}$ in. How much rain fell in all during those three weeks?

____

Use data from the table for problems 3–10.

Rainfall This Year

Month	Inches
January	$2\frac{1}{2}$
February	$2\frac{3}{4}$
March	$4\frac{1}{4}$
April	$4\frac{1}{2}$
May	$2\frac{3}{4}$

3. What was the difference in rainfall between April and May? ________
4. The record for rainfall in March is $9\frac{1}{2}$ inches. How far from the record was the rainfall this March? ________
5. What was the total rainfall during January and February? ________
6. How much more rain fell in April than in January? ________
7. What was the total rainfall during April and May? ________
8. Estimate the total rainfall for all five months. ________
9. How much rain fell in all during the first three months of the year? ________
10. How much rain fell in all during February and March? ________

Spiral Review

11. 3.2 + 5.63 = ________

12. 8.1 − 2.32 = ________

13. 5.65 + 2.4 = ________

14. 7 − 2.83 = ________

Name ________________________________

Explore Adding Fractions with Unlike Denominators

Add.

1. $\frac{1}{2} + \frac{1}{4} =$ ______ **2.** $\frac{1}{3} + \frac{1}{6} =$ ______ **3.** $\frac{1}{4} + \frac{1}{6} =$ ______

4. $\frac{3}{10} + \frac{2}{5} =$ ______ **5.** $\frac{2}{3} + \frac{1}{4} =$ ______ **6.** $\frac{3}{8} + \frac{1}{4} =$ ______

7. $\frac{2}{5} + \frac{1}{4} =$ ______ **8.** $\frac{3}{4} + \frac{1}{6} =$ ______ **9.** $\frac{3}{5} + \frac{3}{10} =$ ______

10. $\frac{3}{4} + \frac{1}{12} =$ ______ **11.** $\frac{3}{8} + \frac{1}{3} =$ ______ **12.** $\frac{1}{6} + \frac{1}{5} =$ ______

13. $\frac{3}{5} + \frac{3}{12} =$ ______ **14.** $\frac{3}{5} + \frac{1}{4} =$ ______ **15.** $\frac{1}{3} + \frac{1}{4} =$ ______

16. $\frac{3}{8} + \frac{1}{2} =$ ______ **17.** $\frac{2}{3} + \frac{1}{6} =$ ______ **18.** $\frac{1}{4} + \frac{1}{8} =$ ______

Solve.

19. On the first day of Lanier's vacation, $\frac{1}{4}$ in. of rain fell. On the second day, $\frac{3}{8}$ in. of rain fell. How much rain fell in all during these two days? Express your answer in inches. ______

20. On Monday, Jayne's stock rose $\frac{3}{8}$ per share. On Tuesday, it rose $\frac{1}{2}$ per share. How much did her stock rise per share in all during the two days? ______

Spiral Review

21. $4.2 \times 10.06 =$ ________ **22.** $1.03 \times 2.74 =$ ________

23. $0.25 \div 5 =$ ________ **24.** $3.2 \div 5 =$ ________

25. $5.82 \times 6.3 =$ ________ **26.** $4.8 \div 6 =$ ________

Name ____________________

Explore Subtracting Fractions with Unlike Denominators

Subtract.

1. $\frac{7}{8} - \frac{1}{2} =$ ______ **2.** $\frac{3}{4} - \frac{5}{8} =$ ______ **3.** $\frac{1}{3} - \frac{1}{6} =$ ______

4. $\frac{2}{3} - \frac{1}{4} =$ ______ **5.** $\frac{6}{8} - \frac{1}{4} =$ ______ **6.** $\frac{2}{3} - \frac{3}{8} =$ ______

7. $\frac{5}{6} - \frac{3}{4} =$ ______ **8.** $\frac{2}{3} - \frac{1}{2} =$ ______ **9.** $\frac{1}{2} - \frac{1}{8} =$ ______

10. $\frac{3}{4} - \frac{2}{3} =$ ______ **11.** $\frac{5}{6} - \frac{2}{3} =$ ______ **12.** $\frac{2}{6} - \frac{1}{8} =$ ______

13. $\frac{11}{20} - \frac{2}{5} =$ ______ **14.** $\frac{7}{10} - \frac{3}{5} =$ ______ **15.** $\frac{3}{4} - \frac{3}{8} =$ ______

16. $\frac{5}{6} - \frac{7}{12} =$ ______ **17.** $\frac{3}{4} - \frac{3}{10} =$ ______ **18.** $\frac{1}{2} - \frac{3}{10} =$ ______

Solve.

19. Marcia is growing plants for a science project. This week one very young plant grew from $\frac{1}{4}$ in. tall to $\frac{7}{8}$ in. tall. How much did the plant grow this week? ______

20. A repair order calls for $\frac{3}{8}$-inch nails. Fred plans to use $\frac{3}{4}$-inch nails. How much longer are the nails Fred plans to use? ______

Spiral Review

Name the value of the underlined digit.

21. $23.\underline{6}25$ ______ **22.** $15.8\underline{7}3$ ______

23. $\underline{3}2.859$ ______ **24.** $15.32\underline{5}$ ______

25. $6\underline{8}.523$ ______ **26.** $51.\underline{5}92$ ______

Name ____________________

Add and Subtract Fractions with Unlike Denominators

Add or subtract. Write your answer in simplest form.

1. $\frac{1}{2} + \frac{1}{3} =$ ______

2. $\frac{3}{5} - \frac{3}{8} =$ ______

3. $\frac{3}{4} - \frac{2}{3} =$ ______

4. $\frac{3}{10} + \frac{1}{4} =$ ______

5. $\frac{1}{2} + \frac{5}{6} =$ ______

6. $\frac{3}{4} + \frac{1}{2} =$ ______

7. $\frac{7}{8} - \frac{1}{4} =$ ______

8. $\frac{5}{6} - \frac{7}{12} =$ ______

9. $\frac{5}{6} + \frac{7}{24} =$ ______

10. $\frac{1}{2} - \frac{1}{5}$

11. $\frac{2}{3} + \frac{5}{6}$

12. $\frac{3}{5} + \frac{7}{10}$

13. $\frac{7}{8} - \frac{2}{3}$

Problem Solving

14. Andy picks a big bag of apples at an orchard. When he gets home, he gives $\frac{1}{2}$ of the apples to his neighbor and another $\frac{1}{4}$ of the apples to his friend. What fraction of his apples does Andy give away? ______

15. It takes Juana $\frac{1}{2}$ hour to drive from her home to Mel's Garage. It takes her another $\frac{1}{6}$ hour to drive from Mel's Garage to Barbara's house. What fraction of an hour does it take for Juana to drive from her home to Barbara's house? ______

Spiral Review

Write whether each number is prime or composite.

16. 43 ______

17. 39 ______

18. 63 ______

19. 54 ______

20. 31 ______

21. 19 ______

Name ______________________

Explore Adding Mixed Numbers with Unlike Denominators

Add. Write your answer in simplest form.

1. $2\frac{1}{3} + 4\frac{5}{6}$ = ______
2. $6\frac{1}{3} + 4\frac{3}{4}$ = ______
3. $7\frac{1}{2} + 3\frac{1}{4}$ = ______
4. $2\frac{7}{10} + 4\frac{2}{5}$ = ______
5. $5\frac{7}{10} + 8\frac{1}{2}$ = ______
6. $10\frac{1}{2} + 2\frac{5}{6}$ = ______
7. $11\frac{7}{12} + 15\frac{2}{3}$ = ______
8. $13\frac{3}{5} + 3\frac{3}{4}$ = ______
9. $17\frac{7}{8} + 6\frac{2}{5}$ = ______
10. $19\frac{5}{6} + 8\frac{3}{4}$ = ______
11. $13\frac{5}{6} + 18\frac{3}{4}$ = ______
12. $22\frac{2}{3} + 26\frac{5}{6}$ = ______

Solve.

13. Harold owns stock in an insurance company. On Monday the stock was worth $16\frac{1}{2}$ per share. On Tuesday it rose in value by $4\frac{3}{4}$. What was the stock worth at the end of the day on Tuesday? ______

14. Fargo, ND, had snowstorms on two days in a row in March. The first storm left $8\frac{3}{4}$ in. of snow, and the second storm left $12\frac{3}{4}$ in. of snow. How many inches of snow did the two storms leave in all? ______

Spiral Review

15. 4.7 + 2.6 = ______
16. 7.5 ÷ 1.5 = ______
17. 3.33 ÷ 3 = ______
18. 25 ÷ 2.5 = ______

Name ___

Add Mixed Numbers with Unlike Denominators

Add. Write your answer in simplest form.

1. $4\frac{2}{3} + 2\frac{1}{6} =$ ___

2. $5\frac{3}{4} + 2\frac{1}{2} =$ ___

3. $8\frac{7}{8} + 4 =$ ___

4. $6\frac{2}{3} + 4\frac{2}{3} =$ ___

5. $5\frac{3}{20} + 24\frac{1}{4} =$ ___

6. $\frac{3}{5} + 4\frac{3}{10} =$ ___

7. $6\frac{1}{2} + 5\frac{3}{4} =$ ___

8. $3\frac{7}{12} + 18\frac{3}{4} =$ ___

9. $4\frac{4}{5} + 3\frac{3}{20} =$ ___

10. $14\frac{1}{2} + 12\frac{2}{3}$

11. $18\frac{5}{6} + 12\frac{2}{3}$

12. $35\frac{4}{5} + 19\frac{3}{5}$

13. $12\frac{11}{12} + 6\frac{1}{6}$

Problem Solving

14. Two boards are being cut into pieces to make a garden border. One is $8\frac{1}{2}$ feet long, and the other is $5\frac{3}{4}$ feet long. If both boards are completely used, how long will the border be? ___

15. Sally delivers newspapers once a week in the morning before school for $1\frac{1}{2}$ hours and after school for $2\frac{1}{2}$ hours. How long does she work on the day she delivers papers? ___

Spiral Review

Find the LCM for each pair of numbers.

16. 6 and 9 ___

17. 5 and 8 ___

18. 8 and 12 ___

19. 5 and 4 ___

20. 9 and 5 ___

21. 5 and 6 ___

Name ________________________________

6·8 Properties of Addition

Find each missing number. Identify the property you used.

1. $14\frac{2}{3} + \_\_\_\_ = 14\frac{2}{3}$ ________

2. $\_\_ + 3\frac{2}{5} = 3\frac{2}{5} + 4\frac{2}{3}$ ________

3. $\_\_ + \left(\frac{2}{5} + \frac{2}{3}\right) = \left(\frac{3}{4} + \frac{2}{5}\right) + \frac{2}{3}$ ________

4. $4\frac{1}{2} + 7\frac{1}{3} = 7\frac{1}{3} + \_\_$ ________

Use the Associative Property to solve. Show your work.

5. $\frac{1}{4} + \left(\frac{1}{4} + \frac{1}{2}\right) =$ ________

6. $2\frac{1}{2} + \left(4\frac{1}{2} + \frac{5}{8}\right) =$ ________

7. $\left(\frac{3}{5} + \frac{1}{6}\right) + \frac{5}{6} =$ ________

8. $\frac{7}{10} + \left(\frac{3}{10} + \frac{2}{3}\right) =$ ________

Problem Solving

9. Carmen has a job walking dogs. On Saturday she worked for $2\frac{1}{2}$ hours, and on Sunday she worked for $3\frac{3}{4}$ hours. How many hours did she work in all during the weekend? ________

10. A gift box needs two lengths of ribbon to be properly wrapped. One length must be $3\frac{1}{2}$ inches long, and the other must be $8\frac{3}{4}$ inches long. What is the total length of ribbon needed to wrap the box? ________

Spiral Review

11. 4(3 + 5) = ________

12. 6(2 + ________) = 24

13. 5(________ – 4) = 15

14. ________(2 + 3) = 25

Name ______________________________

Problem Solving: Strategy

Write an Equation

Write an equation, then solve.

1. Jim's stock started the day at $12\frac{3}{8}$ per share and rose to $15\frac{3}{4}$ per share. How much did the stock increase per share? ______________

2. One weekend Stu worked $7\frac{1}{2}$ hours on Saturday and $6\frac{3}{4}$ hours on Sunday. How many hours did he work in all during the weekend? ______________

3. Karen's grandmother gave her a $2\frac{1}{2}$-pound tin of cookies. Karen brought $1\frac{1}{4}$ pounds of cookies to school to share with her class. How many pounds of cookies does that leave? ______________

4. Latisha's puppy weighed $3\frac{1}{4}$ pounds when they brought her home. Now the puppy weighs $8\frac{7}{8}$ pounds. How many pounds did the puppy gain? ______________

5. At the start of the school year, Jayne was $47\frac{3}{4}$ in. tall. During the year she grew $2\frac{3}{8}$ in. What is her height now? ______________

Spiral Review

Find the median.

6. 3, 6, 2, 9, 15 ________
7. 12, 18, 22, 8 ________
8. 12, 19, 6, 16, 18, 18, 13 ________
9. 8, 9, 3, 15, 21, 12 ________

Name ____

Explore Subtracting Mixed Numbers with Unlike Denominators

Subtract. Write your answer in simplest form.

1. $7\frac{1}{2} - 3\frac{1}{3}$

2. $4\frac{1}{2} - 1\frac{2}{3}$

3. $5\frac{1}{4} - 4\frac{5}{6}$

4. $6\frac{3}{8} - 4\frac{5}{6}$

5. $4\frac{1}{2} - 3\frac{1}{2}$

6. $4\frac{2}{3} - 1\frac{7}{8}$

7. $14\frac{3}{4} - 6\frac{1}{2}$

8. $7\frac{7}{16} - 3\frac{1}{4}$

9. $13\frac{3}{8} - 5\frac{3}{4}$

10. $7\frac{1}{2} - 3\frac{5}{6}$

11. $7\frac{2}{3} - 4\frac{2}{3}$

12. $7\frac{1}{4} - 4\frac{1}{2}$

13. $4\frac{2}{3} - 2\frac{3}{4}$

14. $4\frac{3}{4} - 1\frac{3}{4}$

15. $5\frac{1}{3} - 1\frac{5}{6}$

16. $7\frac{2}{3} - 1\frac{1}{4}$

Solve.

17. A can of tuna fish contains $6\frac{1}{2}$ ounces of tuna. Harry used $2\frac{3}{4}$ ounces for his sandwich. How much tuna fish is left? ________

18. Cory's hat size is $5\frac{1}{2}$. His father's hat size is 8. How much larger is his father's hat size? ________

Spiral Review

Compare. Write <, >, or =.

19. 1.4 ________ $1\frac{2}{5}$

20. $2\frac{3}{4}$ ________ 2.8

21. 5.6 ________ $5\frac{1}{2}$

Name ____________________

6•11 Subtract Mixed Numbers

Subtract. Write your answer in simplest form.

1. $7\frac{1}{8} - 2\frac{3}{4}$

2. $5\frac{1}{2} - 2\frac{5}{6}$

3. $4\frac{2}{3} - 1\frac{5}{6}$

4. $10\frac{1}{3} - 4\frac{7}{10}$

5. $8 - 2\frac{2}{3}$

6. $83\frac{4}{5} - 5\frac{1}{4}$

7. $6\frac{5}{6} - 2\frac{1}{4}$

8. $5\frac{3}{8} - 2\frac{3}{8}$

9. $18\frac{7}{8} - 5\frac{1}{3}$

10. $5\frac{3}{5} - 3$

11. $9\frac{1}{2} - 4\frac{2}{3}$

12. $7\frac{11}{20} - 2\frac{1}{5}$

Problem Solving

13. Kendra is baking an apple pie. The recipe calls for $3\frac{2}{3}$ pounds of apples. Kendra has a 5-pound bag of apples. How many pounds of apples will be left over? ______

14. A stock that Matt owns fell in value from $83\frac{3}{8}$ to $77\frac{1}{2}$. How much did the stock fall? ______

Spiral Review

Simplify each fraction.

15. $\frac{36}{48}$ ______

16. $\frac{40}{56}$ ______

17. $\frac{32}{48}$ ______

18. $\frac{18}{32}$ ______

Name ____________________

Estimate Sums and Differences of Mixed Numbers

Round to the nearest whole number.

1. $8\frac{1}{3}$ ______ 2. $5\frac{4}{5}$ ______ 3. $7\frac{2}{7}$ ______ 4. $54\frac{3}{4}$ ______

Estimate.

5. $4\frac{2}{3} + 7\frac{1}{5}$ ______
6. $8\frac{2}{3} - 4\frac{1}{4}$ ______
7. $8\frac{1}{3} + 6\frac{1}{4}$ ______
8. $9\frac{2}{3} - 8\frac{1}{5}$ ______
9. $8\frac{1}{4} + 4\frac{7}{8}$ ______
10. $6\frac{3}{4} - 2\frac{1}{3}$ ______
11. $9\frac{2}{3} + 8\frac{5}{6}$ ______
12. $20\frac{1}{5} - 17\frac{3}{8}$ ______
13. $21\frac{3}{5} - 17\frac{7}{8}$ ______
14. $9\frac{3}{5} + 8\frac{1}{4}$ ______
15. $3\frac{1}{4} + 18\frac{2}{9}$ ______
16. $15\frac{3}{4} - 7\frac{1}{4}$ ______
17. $18\frac{7}{8} - 4\frac{3}{5}$ ______
18. $8\frac{9}{15} - 7\frac{1}{20}$ ______
19. $3\frac{3}{7} + 5\frac{1}{5}$ ______
20. $19\frac{7}{8} - 15\frac{2}{3}$ ______

Problem Solving

21. Two pieces of string are needed to tie up a package for mailing. One piece must be $15\frac{7}{8}$ in. long, and the other piece must be $22\frac{1}{4}$ inches long. How much string is needed to tie up the package? ______

22. One side of a square picture is $5\frac{1}{3}$ inches long. How much framing is needed for all four sides of the picture? ______

Spiral Review

For each pair of numbers, find the GCF.

23. 24 and 26 ______ 24. 32 and 24 ______

25. 10 and 25 ______ 26. 16 and 28 ______

Name ________________________________

Multiply a Whole Number by a Fraction

Multiply.

1. $\frac{2}{3} \times 48 =$ ______ **2.** $\frac{1}{5} \times 20 =$ ______ **3.** $\frac{3}{4} \times 20 =$ ______

4. $\frac{5}{6} \times 12 =$ ______ **5.** $\frac{1}{3} \times 33 =$ ______ **6.** $40 \times \frac{3}{5} =$ ______

7. $\frac{5}{8} \times 16 =$ ______ **8.** $\frac{9}{10} \times 80 =$ ______ **9.** $\frac{2}{5} \times 350 =$ ______

10. $\frac{1}{4} \times 24 =$ ______ **11.** $\frac{2}{3} \times 45 =$ ______ **12.** $\frac{3}{8} \times 56 =$ ______

13. $\frac{1}{2} \times 38 =$ ______ **14.** $36 \times \frac{5}{12} =$ ______ **15.** $\frac{4}{5} \times 35 =$ ______

16. $\frac{7}{10} \times 60 =$ ______ **17.** $\frac{3}{4} \times 36 =$ ______ **18.** $\frac{2}{3} \times 24 =$ ______

19. $\frac{3}{8} \times 24 =$ ______ **20.** $27 \times \frac{2}{9} =$ ______ **21.** $\frac{3}{8} \times 16 =$ ______

Problem Solving

22. Ben and his family are taking a 450-mile car trip to visit relatives. After they have traveled $\frac{2}{5}$ of the way, they stop for lunch. How many miles have they traveled so far? ______

23. Two-thirds of Paul's class have signed up to play soccer. There are 24 students in the class. How many have signed up for soccer? ______

Spiral Review

24. $\frac{2}{3} + \frac{1}{5} =$ ______ **25.** $4\frac{1}{2} + 2\frac{2}{3} =$ ______

26. $\frac{5}{6} + \frac{3}{8} =$ ______ **27.** $5\frac{1}{3} + 4\frac{1}{12} =$ ______

Name ______________________________

Multiply a Fraction by a Fraction

Multiply. Write each answer in simplest form.

1. $\frac{2}{5} \times \frac{1}{4} =$ ______
2. $\frac{3}{5} \times \frac{2}{9} =$ ______
3. $\frac{3}{4} \times \frac{5}{7} =$ ______
4. $\frac{5}{6} \times \frac{3}{10} =$ ______
5. $\frac{2}{3} \times \frac{4}{5} =$ ______
6. $\frac{3}{8} \times \frac{4}{5} =$ ______
7. $\frac{1}{2} \times \frac{3}{5} =$ ______
8. $\frac{3}{16} \times \frac{4}{5} =$ ______
9. $\frac{3}{8} \times \frac{5}{6} =$ ______
10. $\frac{1}{3} \times \frac{1}{6} =$ ______
11. $\frac{2}{3} \times \frac{3}{5} =$ ______
12. $\frac{3}{8} \times \frac{1}{6} =$ ______
13. $\frac{7}{12} \times \frac{3}{5} =$ ______
14. $\frac{3}{8} \times \frac{4}{7} =$ ______
15. $\frac{1}{4} \times \frac{1}{5} =$ ______
16. $\frac{2}{5} \times \frac{3}{8} =$ ______
17. $\frac{3}{4} \times \frac{5}{12} =$ ______
18. $\frac{5}{12} \times \frac{3}{10} =$ ______
19. $\frac{5}{12} \times \frac{3}{5} =$ ______
20. $\frac{5}{8} \times \frac{3}{10} =$ ______
21. $\frac{2}{5} \times \frac{5}{6} =$ ______

Problem Solving

22. One-half of the animals at the Middletown Zoo are mammals. One-sixth of the mammals belong to the cat family. What fraction of the animals at the zoo belong to the cat family? ______

23. Ryan spends $\frac{1}{2}$ of his homework time working on the computer.

    He spends $\frac{1}{4}$ of his computer time connected to the Internet.

    What fraction of his homework time does he spend connected to the Internet? ______

Spiral Review

Compare. Write <, >, or =.

24. 0.25 ______ $\frac{1}{5}$
25. $\frac{2}{3}$ ______ $\frac{3}{5}$
26. 0.6 ______ $\frac{2}{3}$
27. 0.75 ______ $\frac{3}{4}$

Name ______________________________

Problem Solving: Reading for Math

Solve Multistep Problems

Solve. Write the inference that you made.

1. What length of molding is needed to make a picture frame that is $4\frac{1}{2}$ in. tall and $8\frac{1}{2}$ in. wide?

____

2. Amanda can walk to the park in $\frac{3}{4}$ of an hour. How long will it take her to walk to the park and back?

____

3. Carrie takes $1\frac{3}{4}$ hours to sew a uniform for a toy soldier. How long will it take her to sew 4 uniforms?

____

Solve.

4. A recipe for a single peach pie calls for $2\frac{3}{4}$ pounds of peaches. How many pounds of peaches are needed to make 5 pies? ______

5. Freida uses $2\frac{1}{2}$ yards of cloth to make 2 tablecloths. How many yards of cloth will she need to make 6 tablecloths? ______

6. Art's pancake recipe calls for $1\frac{3}{4}$ cups of flour to make 3 servings. How much flour will Art need to serve 9 people? ______

Spiral Review

7. 1.3×2.73 ______

8. $45 \div 0.05$ ______

9. $3 - 1.634$ ______

10. $100.6 \div 0.02$ ______

Name ________________________________

Multiply Mixed Numbers

Multiply. Write each answer in simplest form.

1. $6\frac{2}{5} \times 2\frac{3}{8} =$ ______ **2.** $4\frac{1}{4} \times 1\frac{1}{3} =$ ______ **3.** $4\frac{1}{6} \times 1\frac{4}{5} =$ ______

4. $8\frac{1}{4} \times 2\frac{2}{3} =$ ______ **5.** $\frac{2}{3} \times 12\frac{1}{2} =$ ______ **6.** $2\frac{1}{2} \times \frac{3}{4} =$ ______

7. $6\frac{2}{3} \times 2\frac{7}{10} =$ ______ **8.** $4\frac{1}{3} \times 5\frac{2}{5} =$ ______ **9.** $3\frac{1}{3} \times 10 =$ ______

10. $8\frac{1}{3} \times 4\frac{4}{5} =$ ______ **11.** $6\frac{1}{2} \times 5\frac{1}{3} =$ ______ **12.** $\frac{1}{15} \times 2\frac{1}{2} =$ ______

13. $2\frac{7}{10} \times 3\frac{1}{3} =$ ______ **14.** $\frac{3}{5} \times 2\frac{1}{6} =$ ______ **15.** $7\frac{1}{3} \times 9\frac{1}{2} =$ ______

16. $12\frac{1}{2} \times 2\frac{1}{2} =$ ______ **17.** $4\frac{1}{5} \times 2\frac{1}{7} =$ ______ **18.** $8\frac{4}{5} \times 2\frac{3}{11} =$ ______

19. $4\frac{1}{2} \times 2\frac{2}{3} =$ ______ **20.** $7\frac{1}{3} \times 2\frac{5}{11} =$ ______ **21.** $4\frac{2}{5} \times 3\frac{3}{4} =$ ______

Problem Solving

22. Kate carves dolls from lengths of plywood. Each doll is $2\frac{3}{4}$ inches tall. How many inches of plywood does she need for 8 dolls? ______________

23. Carole measures her height in $4\frac{3}{4}$-inch clothespins. She is $12\frac{1}{2}$ clothespins tall. How many inches tall is she? ______________

Spiral Review

Label as prime or composite.

24. 37 ______________ **25.** 27 ______________

26. 57 ______________ **27.** 67 ______________

Name____

Estimate Products

Estimate.

1. $53 \times \frac{1}{5}$ ______ 2. $45\frac{1}{5} \times \frac{3}{4}$ ______ 3. $5\frac{3}{5} \times 2\frac{1}{12}$ ______

4. $28\frac{1}{3} \times \frac{2}{3}$ ______ 5. $4\frac{4}{5} \times 5\frac{1}{4}$ ______ 6. $9\frac{3}{5} \times 2\frac{1}{8}$ ______

7. $8\frac{5}{8} \times \frac{2}{3}$ ______ 8. $3\frac{2}{7} \times 2\frac{3}{10}$ ______ 9. $31\frac{1}{8} \times \frac{2}{3}$ ______

10. $1\frac{2}{3} \times 4\frac{3}{5}$ ______ 11. $7\frac{1}{3} \times 8\frac{1}{6}$ ______ 12. $16\frac{5}{7} \times 4$ ______

13. $\frac{2}{3} \times 20\frac{4}{5}$ ______ 14. $\frac{1}{5} \times 25\frac{1}{12}$ ______ 15. $6\frac{1}{3} \times 2\frac{1}{3}$ ______

Estimate to compare. Write >, <, or =.

16. $4\frac{1}{4} \times 3\frac{1}{8}$ ______ $3\frac{8}{9} \times 2\frac{5}{6}$ 17. $9\frac{2}{7} \times 1\frac{15}{16}$ ______ $3\frac{2}{9} \times 6\frac{9}{10}$

18. $2\frac{4}{5} \times 5\frac{1}{12}$ ______ $4\frac{1}{9} \times 3\frac{7}{8}$ 19. $8\frac{1}{5} \times 3\frac{7}{8}$ ______ $4\frac{11}{12} \times 5\frac{2}{3}$

Problem Solving

Estimate the solutions.

20. Kareem delivers newspapers every day. It takes him about $1\frac{3}{4}$ hours per day to do the job. About how many hours a week does he spend delivering papers? ______

21. Freida makes pots for plants and sells them in sets of three. If each pot weighs $2\frac{1}{8}$ pounds, about how much does a three-pot set weigh? ______

Spiral Review

Round to the underlined digit.

22. $3.\underline{2}45$ ______ 23. $8.6\underline{7}6$ ______ 24. $\underline{8}.732$ ______

Name ______________________________

7·6 Problem Solving: Strategy

Make an Organized List

Make an organized list to solve.

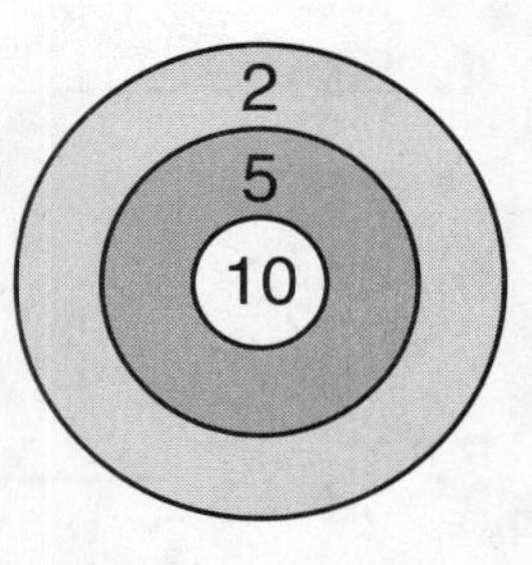

1. Kurt and Matthew play darts using the dartboard shown. The numbers indicate the points scored for hitting each area on the board. If Kurt and Matthew each throw one dart and both hit some part of the board, what are their possible combined scores?

____

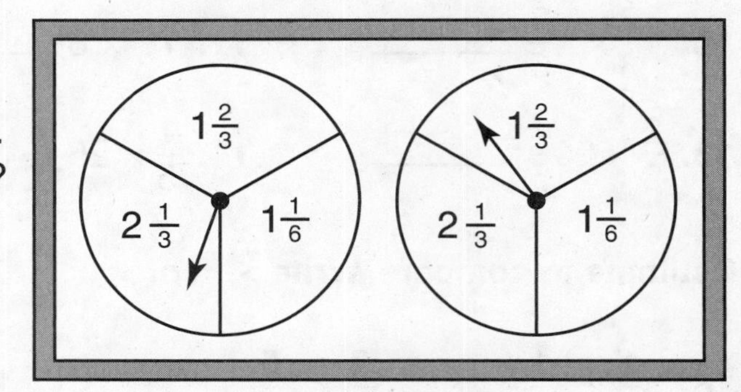

2. Francine plays this spinner game. She spins both spinners and finds the products of the mixed numbers. What products can Francine make?

____

Mixed Strategy Review

3. Six students stand in line for lunch. Hank is first in line and Gwen is fifth. Keisha is not standing next to Hank or next to Gwen. Latrelle is standing just ahead of Keisha. Janice is not fourth in line. Where in the line is Fred?

____

4. Maria has collected 16 oak leaves and 18 maple leaves. Chuck has collected $\frac{1}{4}$ as many oak leaves and $1\frac{1}{3}$ as many maple leaves. How many leaves has Chuck collected? ______________________________

Spiral Review

5. $\frac{3}{5} - \frac{1}{10} =$ _____

6. $4 - 2\frac{1}{2} =$ _____

7. $5\frac{3}{4} - 2\frac{1}{6} =$ _____

8. $7\frac{1}{2} - 3\frac{2}{3} =$ _____

Name ____

Properties of Multiplication

Identify each property of multiplication.

1. $4(3 + 5) = 4 \times 3 + 4 \times 5$

____

2. $8\frac{1}{3} \times 4\frac{1}{2} = 4\frac{1}{2} \times 8\frac{1}{3}$

____

3. $6\frac{2}{3} \times 1 = 6\frac{2}{3}$

____

4. $18\frac{2}{3} \times 0 = 0$

____

5. $7\frac{2}{3} \times \left(4\frac{1}{2} \times 6\right) = \left(7\frac{2}{3} \times 4\frac{1}{2}\right) \times 6$

____

6. $2\frac{2}{3} \times 4\frac{4}{5} = 4\frac{4}{5} \times 2\frac{2}{3}$

____

Use a multiplication property to solve. Show your work.

7. $\left(3\frac{1}{2} \times 2\frac{2}{3}\right) \times 6 =$

____

8. $7\frac{3}{4} \times 8 =$

____

Problem Solving

9. Gwen has a three-hour phone card. She uses $1\frac{1}{3}$ hours to call her sister and $\frac{3}{4}$ of an hour to call her aunt. How much time is left on her card?

____

10. Ken has 10 pounds of plaster. He plans to make five figurines using $1\frac{3}{4}$ pounds of plaster for each one. How much plaster will Ken have left? ________

Spiral Review

Order from least to greatest.

11. $3\frac{1}{3}$, $3\frac{1}{2}$, 3.4 ____________

12. 7.2, $7\frac{1}{4}$, $7\frac{1}{10}$ ____________

Name________________________________

7·8 Explore Dividing Fractions

Divide.

1. $3 \div \frac{1}{6} =$ ______

2. $5 \div \frac{1}{4} =$ ______

3. $8 \div \frac{1}{3} =$ ______

4. $13 \div \frac{1}{5} =$ ______

5. $4 \div \frac{1}{9} =$ ______

6. $7 \div \frac{1}{7} =$ ______

7. $8 \div \frac{1}{8} =$ ______

8. $2 \div \frac{1}{2} =$ ______

9. $5 \div \frac{1}{3} =$ ______

10. $6 \div \frac{1}{5} =$ ______

11. $6 \div \frac{1}{4} =$ ______

12. $7 \div \frac{1}{8} =$ ______

13. $4 \div \frac{1}{6} =$ ______

14. $3 \div \frac{1}{4} =$ ______

15. $13 \div \frac{1}{3} =$ ______

16. $17 \div \frac{1}{10} =$ ______

17. $8 \div \frac{1}{12} =$ ______

18. $36 \div \frac{1}{10} =$ ______

19. $15 \div \frac{1}{3} =$ ______

20. $6 \div \frac{1}{8} =$ ______

21. $4 \div \frac{1}{12} =$ ______

Solve.

22. Each of Conan's chocolates weighs $\frac{1}{3}$ of an ounce. How many chocolates are there in a 6-ounce box? ______

23. Harry is cutting $\frac{1}{2}$-inch blocks from a 12-inch wooden board. How many blocks can he cut? ______

Spiral Review

Give the GCF.

24. 12, 20 ______

25. 24, 30 ______

26. 18, 42 ______

27. 24, 36 ______

Name ______________________________

Divide Fractions

Write the reciprocal of each number.

1. $\frac{2}{3}$ ____ **2.** $\frac{5}{6}$ ____ **3.** 6 ____ **4.** $2\frac{3}{4}$ ____

Divide. Write each answer in simplest form.

5. $4 \div \frac{2}{3} =$ 6

6. $\frac{1}{6} \div \frac{2}{3} =$ ____

7. $\frac{3}{8} \div \frac{1}{4} =$ ____

8. $\frac{1}{2} \div \frac{1}{6} =$ ____

9. $\frac{7}{10} \div \frac{3}{5} =$ ____

10. $\frac{2}{5} \div \frac{3}{8} =$ ____

11. $\frac{7}{12} \div \frac{3}{4} =$ ____

12. $\frac{8}{15} \div \frac{2}{5} =$ ____

13. $\frac{3}{5} \div \frac{2}{3} =$ ____

14. $\frac{2}{3} \div \frac{4}{5} =$ ____

15. $\frac{3}{5} \div \frac{3}{5} =$ ____

16. $\frac{1}{4} \div \frac{2}{7} =$ ____

Problem Solving

17. Martina is meeting her friends at the mall. They will meet at a bench located on the mall's $\frac{3}{4}$-mile walker's route. The bench marks the $\frac{1}{3}$ point on the route. If Martina begins at the start of the route, how far must she walk to meet her friends? ________

18. A standard sheet of lined paper is $8\frac{1}{2}$ inches wide. How many $\frac{1}{2}$-inch columns can be drawn on the paper? ________

Spiral Review

Give the LCM.

19. 6, 8 ____ **20.** 3, 6 ____ **21.** 2, 5 ____

22. 4, 6 ____ **23.** 4, 3 ____ **24.** 6, 9 ____

Name ____________________

Divide Mixed Numbers

Divide. Write each answer in simplest form.

1. $8\frac{1}{4} \div 3\frac{2}{3} =$ ______

2. $6\frac{2}{3} \div \frac{2}{3} =$ ______

3. $4\frac{1}{2} \div \frac{3}{4} =$ ______

4. $4\frac{3}{4} \div 2 =$ ______

5. $4\frac{2}{3} \div \frac{7}{12} =$ ______

6. $1\frac{3}{10} \div \frac{1}{2} =$ ______

7. $7\frac{3}{4} \div 3\frac{1}{2} =$ ______

8. $3\frac{1}{4} \div 1\frac{3}{10} =$ ______

9. $3\frac{1}{6} \div 6\frac{1}{3} =$ ______

10. $3\frac{1}{2} \div \frac{14}{15} =$ ______

11. $9\frac{3}{5} \div 12 =$ ______

12. $\frac{5}{12} \div 3\frac{1}{3} =$ ______

13. $\frac{4}{5} \div 3\frac{1}{10}$ ______

14. $6\frac{1}{3} \div 3 =$ ______

15. $4\frac{2}{3} \div \frac{1}{6} =$ ______

16. $5\frac{3}{4} \div \frac{7}{8} =$ ______

17. $1\frac{5}{12} \div \frac{1}{12} =$ ______

18. $10\frac{1}{2} \div 4\frac{2}{3} =$ ______

Solve.

19. $4\frac{1}{2} \div n = 9$ ______

20. $5\frac{1}{4} \div n = \frac{3}{8}$ ______

Problem Solving

21. Acme clothespins are $3\frac{1}{2}$ inches long. How many wooden clothespins can be cut from a 21-inch piece of stock? ______

22. Gary sells cookies at his school's fund-raiser. Each package of cookies weighs $2\frac{3}{4}$ ounces. How many packages can he make from 22 ounces of cookies? ______

Spiral Review

Find the mean.

23. 4, 6, 3, 11 ______

24. 8, 15, 12, 20, 25 ______

25. 8, 6, 15, 20 ______

26. 18, 22, 28, 32, 49 ______

Name ____

Time

Complete.

1. 49 d = ________ w **2.** 216 h = ________ d **3.** 730 d = ________ y

4. 2 d 5 h = ________ h **5.** 1 h 15 min = ________ min

6. 50 mo = ________ y ________ mo **7.** 4 d 4 h = ________ h

8. 48 min = ________ s **9.** 8 decades = ________ y

Find each elapsed time.

10. 2:15 A.M. to 8:00 A.M.

____

11. 9:40 P.M. to 11:05 P.M.

____

12. 11:06 A.M. to 3:15 P.M.

____

13. 5:30 A.M. to 11:04 A.M.

____

Find each time.

14. 1 h 30 min after 7:00 A.M. ________________

15. 3 h 43 min after 10:30 A.M. ________________

Problem Solving

16. Mary starts working in her garden one morning at 10:30 A.M. At 1:15 P.M. she goes inside for lunch. How long does she work in her garden?

____

17. Carlo plans to walk across town to a friend's house. He figures it will take him $2\frac{1}{2}$ hours to make the trip. If he wants to get there at 2:45 P.M., when should Carlo leave home?

____

Spiral Review

18. $16 \times \frac{1}{4} =$ ________ **19.** $84 \times \frac{1}{12} =$ ________

20. $24 \div \frac{2}{3} =$ ________ **21.** $30 \div \frac{5}{6} =$ ________

Name ____________________

8•2 Customary Length

Choose an appropriate unit to measure the length of each. Write *in., ft, yd,* or *mi.*

1. height of a table ________

2. distance across a baseball field ________

3. width of a book ________

4. distance from New York to Chicago ________

Complete.

5. 4 yd = ______ in.

6. 100 in. = _____ ft _____ in.

7. 18 yd = _____ ft

8. 75 ft = _____ yd

9. 180 in. = _____ yd

10. 30 ft = _____ yd

11. 6 ft = _____ in.

12. 108 ft = _____ yd

13. 3 ft 6 in. = _____ in.

14. 4 ft 10 in. = _____ in.

15. 8 yd = _____ in.

Compare. Write <, >, or =.

16. 9 ft _____ 4 yd

17. 150 in. _____ 10 ft

18. 345 ft _____ 115 yd

19. 180 in. _____ 5 ft

20. 11 yd _____ 500 in.

21. 35 in. _____ 2 ft 10 in.

Problem Solving

22. Kandy's height is 5 feet 1 inch. What is Kandy's height in inches? ________

23. A football field is 100 yards long. How long is this in feet? ________

How long is this in inches? ________

Spiral Review

24. $4\frac{1}{2} \times \frac{7}{8} =$ ______

25. $5\frac{3}{4} \times 6\frac{2}{3} =$ ______

26. $2\frac{5}{8} \times 5\frac{1}{7} =$ ______

27. $16 \times 4\frac{7}{8} =$ ______

Name ______________________________

Customary Capacity and Weight

Choose an appropriate unit to measure the capacity of each. Write *fl oz, c, pt, qt,* or *gal.*

1. juice box ______________ **2.** water pitcher ______________

3. home aquarium ______________

Choose an appropriate unit to measure the weight of each. Write *oz, lb,* or *T.*

4. newborn kitten ________ **5.** truck ________ **6.** bag of apples ________

Complete.

7. 3 c = ______ fl oz

8. 20 pt = ______ c

9. 12 pt = ______ qt

10. 30 fl oz = ______ c ______ fl oz

11. 7 gal = ______ qt

12. 6 qt = ______ pt

13. 6 lb = ______ oz

14. $2\frac{1}{2}$ T = ______ lb

15. 24 qt = ______ gal

16. 80 oz = ______ lb

17. 80 fl oz = ______ qt ______ pt

18. 10,000 lb = ______ T

Compare. Write >, <, or =.

19. 26 c ______ 200 fl oz **20.** 16 lb ______ 256 oz **21.** 35 pt ______ 17 qt

Problem Solving

22. A juice box contains 4 fl oz. How many juice boxes of orange juice can be poured into a 1-gallon pitcher? ______________

23. Ariel's family is having a cookout. They plan to serve hamburgers. A typical hamburger patty weighs 4 oz. How many patties can be made from a 3-pound package of chopped meat? ______________

Spiral Review

Find the LCD.

24. $\frac{1}{6}, \frac{1}{4}$ ________ **25.** $\frac{4}{5}, \frac{2}{3}$ ________ **26.** $\frac{3}{8}, \frac{3}{10}$ ________

Name ______________________________

Problem Solving: Reading for Math
Check for Reasonableness

Is each estimate reasonable? Explain.

1. Jack's family just bought a new car. Jack says it is about 50 feet long.

____

2. Jack also says his family's new car weighs about 400 pounds.

____

3. Mary's house is 30 feet wide. She says this is 90 yards.

____

____

4. Karen lives two miles from her school. She says this distance is about 10,000 feet.

____

5. A swimming pool at the town recreation center is about 9 feet deep at one end. Nick says this is about 100 inches.

____

6. Jan needs 12 feet of rope for a jump rope competition. She asks her mother to buy 36 yards of rope.

____

____

7. In the owner's manual for his lawn mower, Harry reads that he needs 16 fluid ounces of oil. He buys 1 quart of oil.

____

____

Spiral Review

Solve.

8. $4(3 + 2) =$ ______

9. $5(6 +$ ______ $) = 45$

10. $7($ ______ $+ 3) = 35$

11. ______ $(4 + 6) = 80$

Name ______________________________

Explore Metric Length

Choose an appropriate unit. Write *mm, cm, m,* or *km.*

1. length of your math book ____________
2. your height ____________
3. width of your classroom ____________
4. thickness of a crayon ____________
5. distance from your school to home ____________
6. length of the school playground ____________
7. height of the classroom ceiling ____________
8. length of your bed ____________
9. thickness of 10 pieces of paper ____________
10. length of a dollar bill ____________
11. width of a quarter ____________
12. distance across your town ____________

Solve.

13. Would you measure the width of a standard television screen in centimeters or meters? Explain.

____

____

14. Would you measure the height of your school in meters or kilometers? Explain.

____

____

Spiral Review

15. $\frac{7}{9} - \frac{1}{3} =$ ______ 16. $4 - 2\frac{1}{2} =$ ______ 17. $6\frac{3}{4} - 5\frac{1}{2} =$ ______

Name ______________________________

Metric Capacity and Mass

Choose an appropriate unit of capacity to measure each. Write *mL* or *L*.

1. soda bottle ________ 2. tablespoon ________ 3. thermos ________

4. cereal bowl ________ 5. car gas tank ________ 6. bathtub ________

Choose an appropriate unit of mass to measure each. Write *mg, g,* or *kg*.

7. pencil ________ 8. computer ________

9. 6 sheets of paper ________ 10. bicycle ________

Choose an appropriate estimate for each.

11. small dog ________

A 5 grams **B** 50 grams
C 5 kilograms **D** 50 kilograms

12. CD ________

A 20 milligrams **B** 200 milligrams
C 20 grams **D** 200 grams

13. flower vase ________

A 10 milliliters **B** 1 liter
C 20 liters **D** 100 liters

14. stick of margarine ________

A 250 milliliters **B** 1 liter
C 10 liters **D** 50 liters

Problem Solving

15. Sean is mailing his cousin two boxes of crayons. Do they weigh about 50 grams or about 50 kilograms? ________________

How do you know? ________________

16. Curt is filling up his new fish tank. Will he need about 20 milliliters or about 20 liters of water? ________________

How do you know? ________________

Spiral Review

Write each fraction in simplest form.

17. $\frac{12}{24}$ ____ 18. $\frac{18}{32}$ ____ 19. $\frac{65}{100}$ ____ 20. $\frac{28}{64}$ ____

Name ____

Metric Conversions

Complete.

1. 380 g = _______ kg
2. 180 cm = _______ m
3. 2 L = _______ mL
4. 1,946 cm = _______ m
5. 8 L = _______ cL
6. 13 m = _______ mm
7. 295 m = _______ km
8. 27.4 g = _______ kg
9. 1,853 mm = _______ m
10. 61 cL = _______ L
11. 2,115 mg = _______ g
12. 6,125 mL = _______ cL

Find each sum.

13. 3.2 g + 4 kg + 135.3 g = __________ g
14. 3.2 m + 4.68 m + 123 cm = __________ m
15. 108 m + 225 cm + 4.1 km = __________ m

Compare. Write <, >, or =.

16. 3.6 m ___ 360 cm
17. 8.3 kg ___ 8,300 g
18. 3,685 mL ___ 3.685 L
19. 3.6 km ___ 360 m
20. 236 cm ___ 23.6 m
21. 81 cL ___ 8,100 mL

Problem Solving

22. Sandy packs three gifts in a large box to mail to his cousins. The three gifts weigh 3 kg, 384 g, and 985 g. What is the total shipping weight of the large box in kilograms? ______________

23. Lucy is framing a photograph in the shape of a rectangle. One side of the photograph is 20 cm long and the other side is 135 mm long. In centimeters, what is the length of stock needed for this frame? ______________

Spiral Review

24. 0.3×80 ______________
25. 2.01×0.5 ______________
26. 6×3.2 ______________
27. 0.03×0.03 ______________

Name ______________________________

Problem Solving: Strategy

Draw a Diagram

Draw a diagram to solve.

1. LaTonya's father is building a fence for his garden. The garden is 40 feet by 20 feet. The fence will have fence posts at each corner and also at every 4 feet of fence. How many fence posts will be needed? ________

2. Ginny is going to the movies. She has a ticket for seat number 54. At the theater the first row has 10 seats, and the second row has 9 seats The following rows continue to alternate between 10 and 9 seats. In which row is Ginny's seat? ________

3. Joanie is taking a walk. She begins by heading north. After a while she turns to her left. Later she turns to the opposite direction. Finally, she turns right and then turns right again. In which direction is she now facing? ________

4. Juanita wants to decorate a square picture frame. She wants to put six gold stars on each side, with an additional star at each corner. How many gold stars will she need? ________

Mixed Strategy Review

5. Calvin makes 8 telephone calls on Saturday. This is 4 less than 3 times the number of calls he makes on Sunday. How many telephone calls does Calvin make on Sunday? ________

6. Claire has a piano lesson at 1:15 P.M. The lesson lasts 1 hour 30 minutes. Afterward she reads for 45 minutes, and then she spends 25 minutes doing exercises. At what time does Claire finish doing exercises? ________

Spiral Review

Find the mean.

7. 13, 6, 12, 79, 15 ________

8. 9, 14, 8, 12, 17 ________

9. 6, 14, 12, 20, 22, 25 ________

10. 18, 36, 25, 21, 32, 42, 51 ________

Name ____

8-9 Temperature

Estimate each temperature in °F and °C.

Degrees Fahrenheit (°F)		Degrees Celsius (°C)
212°	Water boils	100°
98.6°	Normal body temperature	37°
68°	Room temperature	20°
32°	Water freezes	0°

1. cool autumn day ____________
2. classroom ____________
3. hot chocolate ____________

Estimate each Fahrenheit temperature.

4. 15°C = ________ °F
5. 18°C = ________ °F
6. 42°C = ________ °F
7. 70°C = ________ °F
8. 23°C = ________ °F
9. 53°C = ________ °F
10. 76°C = ________ °F
11. 62°C = ________ °F
12. 82°C = ________ °F

Problem Solving

13. In Christy's family's new refrigerator, all the temperature gauges are in Celsius. Christy checks the temperature in the freezer. It is 4°C. Should she adjust it? If so, why?

____

____

14. Tricia's school group is traveling to Paris, but she isn't sure what kind of clothes to pack. During the week of the visit, temperatures in Paris are expected to be near 30°C. Should Tricia bring cool clothes or warm clothes? Explain your answer.

____

Spiral Review

15. $4\frac{1}{2} \times 2\frac{2}{3}$ ______
16. $6\frac{1}{8} \div 7$ ______
17. $2\frac{3}{4} \times 1\frac{2}{3}$ ______
18. $2\frac{2}{5} \div \frac{2}{5}$ ______

Name ______________________________

Integers and the Number Line: Comparing and Ordering Integers

Write an integer to represent each situation.

1. A hot air balloon descends 300 feet. __________

2. Michele earns $5.00 selling lemonade. __________

3. Paula took $25.95 out of her bank account. __________

4. An elevator goes from the seventh floor to the third floor. __________

Compare. Write < or >. You may use a number line.

5. 1 _____ $^{-}3$ **6.** $^{-}6$ _____ 2 **7.** $^{-}8$ _____ $^{-}2$ **8.** $^{-}4$ _____ 5

9. 0 _____ $^{-}2$ **10.** $^{-}14$ _____ 13 **11.** $^{-}17$ _____ 13 **12.** 17 _____ $^{-}13$

Order the integers from least to greatest.

13. $^{-}6$, 6, 0 ______________ **14.** 1, $^{-}2$, 3 ______________

15. 5, $^{-}4$, 3 ______________ **16.** $^{-}2$, $^{-}3$, 2 ______________

17. 4, 0, $^{-}3$ ______________ **18.** 1, 0, $^{-}3$ ______________

Problem Solving

19. Marcia and her friends are hiking up a mountain. When they reach the 1,500-foot level, they stop to rest. The next time they stop, they have climbed another 500 feet. How far up the mountain are they when they stop the second time? ______________

20. At the start of the month Billy has $50.63 in his bank account. During the month he takes out $10.40. What is his account balance at the end of the month? ______________

Spiral Review

Find the GCF.

21. 12, 20 __________ **22.** 36, 30 __________

23. 14, 54 __________ **24.** 9, 6 __________

Name ____

Explore Adding Integers

Add.

1. $5 + 2 =$ ________ **2.** $^{-}8 + 3 =$ ________

3. $5 + {}^{-}2 =$ ________ **4.** $7 + {}^{-}8 =$ ________

5. $0 + {}^{-}3 =$ ________ **6.** $7 + 4 =$ ________

7. $0 + {}^{-}9 =$ ________ **8.** $^{-}6 + {}^{-}4 =$ ________

9. $8 + {}^{-}8 =$ ________ **10.** $^{-}6 + 4 =$ ________

11. $5 + {}^{-}5 =$ ________ **12.** $^{-}9 + {}^{-}2 =$ ________

13. $6 + 4 =$ ________ **14.** $9 + {}^{-}3 =$ ________

15. $^{-}8 + 4 =$ ________ **16.** $^{-}12 + 0 =$ ________

17. $^{-}18 + {}^{-}3 =$ ________ **18.** $^{-}9 + 3 =$ ________

19. $6 + {}^{-}6 =$ ________ **20.** $12 - 14 =$ ________

21. $^{-}8 + {}^{-}4 =$ ________ **22.** $^{-}5 + 6 =$ ________

23. $8 + 3 =$ ________ **24.** $16 + {}^{-}9 =$ ________

Solve.

25. On the same day that Sandy withdraws $35 from her bank account, she also deposits a check for $45. Use a signed number to show how her bank account has changed. ____________

26. When Dave reads the thermometer at 4 P.M., the temperature is 63°F. At 6 P.M. the temperature is 58°F. Use a signed number to show how the temperature has changed. ____________

Spiral Review

27. $2.35 \div 0.05 =$ ____________ **28.** $8.34 \times 0.05 =$ ____________

29. $6 - 1.05 =$ ____________ **30.** $6.25 + 10.324 =$ ____________

Name ______________________________

9·3 Add Integers

1. $5 + 7 =$ ______ **2.** $^-7 + 4 =$ ______ **3.** $6 + {}^-3 =$ ______

4. $3 + {}^-7 =$ ______ **5.** $0 + {}^-4 =$ ______ **6.** $3 + 5 =$ ______

7. $0 + {}^-8 =$ ______ **8.** $^-5 + {}^-3 =$ ______ **9.** $5 + {}^-5 =$ ______

10. $^-6 + 8 =$ ______ **11.** $7 + {}^-8 =$ ______ **12.** $^-2 + 2 =$ ______

Complete the function tables.

$y = 2x + {}^-3$

	x	y
13.	$^-6$	
14.	$^-3$	
15.	0	
16.	1	
17.	3	

$y = x + {}^-4$

	x	y
18.	$^-5$	
19.	$^-2$	
20.	0	
21.	3	
22.	6	

Problem Solving

23. An elevator stops at the 54th floor of a skyscraper. Then it rises 5 floors, then descends 8 floors. On what floor is the elevator now? ______

24. A telephone pole is 30 feet long. It is sunk 6 feet into the ground when it is put into service. How high off the ground is the top of the pole? ______

Spiral Review

25. $4\frac{2}{3} + 5\frac{1}{2} =$ ______ **26.** $7\frac{1}{3} - 2\frac{1}{2} =$ ______ **27.** $\frac{5}{8} + 3\frac{2}{5} =$ ______

28. $5 - 2\frac{1}{4} =$ ______ **29.** $6\frac{1}{2} + 5\frac{3}{8} =$ ______ **30.** $12\frac{5}{6} - 7\frac{11}{15} =$ ______

Name ______________________________

Problem Solving: Reading for Math

Check the Reasonableness of an Answer

Use the data from the graph to answer problems 1–5.

1. Claudette knows that air temperatures are normally warmer in sunny spots than in shady spots. At three times during a cool spring day, she takes the temperature of the air at three locations. Then she graphs her results as shown. At the sunny spot, her readings show the temperature rising from 35° in the morning to 50° in the afternoon, then later dipping to 45°. Is this reasonable?

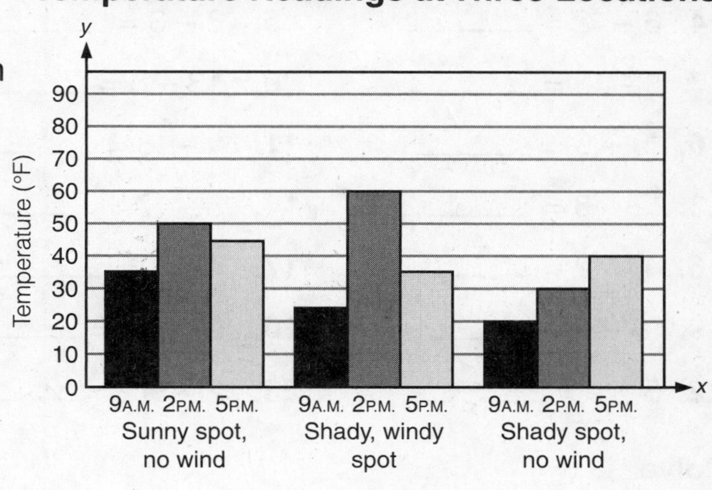

____

2. At the shady, windy spot, her readings show the temperature starting out lower than in the sunny spot, then rising to 60°. Is this reasonable?

____

____

3. At the shady spot with no wind, her readings show the temperature reaching 30° by 2:00 P.M., then rising to 40° by 5:00 P.M. Is this reasonable?

____

____

4. Claudette writes that in the sunny spot, the temperature rose by 15°F between 9:00 A.M. and 2:00 P.M. Is her statement reasonable?

____

Spiral Review

5. 44 mo = ______ y ______ mo

6. 100 min = ______ h ______ min

Name ____________________

Explore Subtracting Integers

Subtract. You may use counters.

1. $^-8 - ^-2 =$ ______ **2.** $5 - ^-2 =$ ______ **3.** $8 - ^-3 =$ ______

4. $6 - ^-2 =$ ______ **5.** $^-8 - 3 =$ ______ **6.** $^-3 - 5 =$ ______

7. $8 - ^-6 =$ ______ **8.** $^-13 - 5 =$ ______ **9.** $7 - ^-2 =$ ______

10. $^-9 - 5 =$ ______ **11.** $^-1 - ^-8 =$ ______ **12.** $7 - ^-7 =$ ______

13. $^-2 - 8 =$ ______ **14.** $^-3 - ^-7 =$ ______ **15.** $1 - ^-9 =$ ______

16. $4 - ^-2 =$ ______ **17.** $^-4 - 7 =$ ______ **18.** $^-6 - ^-6 =$ ______

19. $^-8 - 4 =$ ______ **20.** $^-9 - ^-14 =$ ______ **21.** $1 - ^-6 =$ ______

22. $5 - 9 =$ ______ **23.** $^-2 - ^-12 =$ ______ **24.** $7 - 12 =$ ______

Solve.

25. Construction workers are digging the foundation for a skyscraper. On Monday they reach a level 14 feet below the surface. On Tuesday they dig down another 18 feet. How far below the surface are they now? ______

26. Divers swim down to a reef 42 feet below the ocean surface. Then they swim up 16 feet to investigate the top of a rock ledge. How far below the surface are they now? ______

Spiral Review

Tell whether each number is prime or composite.

27. 33 ______ **28.** 23 ______

29. 63 ______ **30.** 43 ______

31. 93 ______ **32.** 83 ______

Name ______________________________

9·6 Subtract Integers

Subtract.

1. $8 - {}^{-}3 =$ ______ **2.** ${}^{-}3 - {}^{-}3 =$ ______ **3.** ${}^{-}13 - 18 =$ ______

4. ${}^{-}17 - {}^{-}8 =$ ______ **5.** ${}^{-}8 - {}^{-}6 =$ ______ **6.** $7 - {}^{-}2 =$ ______

7. $12 - {}^{-}5 =$ ______ **8.** ${}^{-}15 - {}^{-}5 =$ ______ **9.** $5 - 8 =$ ______

10. ${}^{-}15 - {}^{-}6 =$ ______ **11.** $83 - {}^{-}42 =$ ______ **12.** ${}^{-}8 - 6 =$ ______

13. ${}^{-}17 - 12 =$ ______ **14.** ${}^{-}48 - {}^{-}20 =$ ______ **15.** ${}^{-}6 - 6 =$ ______

16. ${}^{-}12 - {}^{-}15 =$ ______ **17.** $13 - {}^{-}6 =$ ______ **18.** ${}^{-}9 + 4 =$ ______

Complete.

19. ${}^{-}5 - {}^{-}7 = {}^{-}5 +$ ______ $=$ ______ **20.** $12 - {}^{-}6 = 12 +$ ______ $=$ ______

21. $6 -$ ______ $= 6 + {}^{-}9 =$ ______ **22.** $11 -$ ______ $= 11 + 4 =$ ______

Compare. Write <, >, or =.

23. $8 - {}^{-}7$ ______ $18 - 3$ **24.** $2 - 6$ ______ ${}^{-}3 - 4$

25. ${}^{-}16 - {}^{-}7$ ______ $2 - 9$ **26.** $7 - 9$ ______ ${}^{-}6 - {}^{-}3$

Problem Solving

27. An elevator is on the second floor of a building and descends 4 floors to a sub-basement. Write an equation to show where it is now located.

____

28. Weng has $100.32 in his bank account. He withdraws $7.14. Write an equation to show how much money he has now in his account.

____

Spiral Review

Write as a prime factorization.

29. 20 ______ **30.** 36 ______

31. 24 ______ **32.** 30 ______

Name ____________________

Problem Solving: Strategy
Alternate Solution Methods

Solve using two different methods. Tell which methods you used.

1. A fence surrounds a square garden. Each side of the garden is 12 yards long. How long is the fence in feet? ____________

2. Amy is helping her parents seed their yard. Two bags of grass seed will cover about 500 square feet of yard. Amy's yard is about 1,250 square feet. How many bags will she need? ____________

3. Kerry can sell her handmade bracelets to a local craft store for $2.50 each. How many will she have to sell to afford a new CD that costs $15.99? ____________

4. Lamont has a collection of 486 baseball cards. His friend Manuel has a third as many. How many baseball cards do they have in all? ____________

Mixed Strategy Review

Use data from the graph for problems 5–6.

5. Which month had the lowest average daily high temperature?

____

What was that temperature?

____

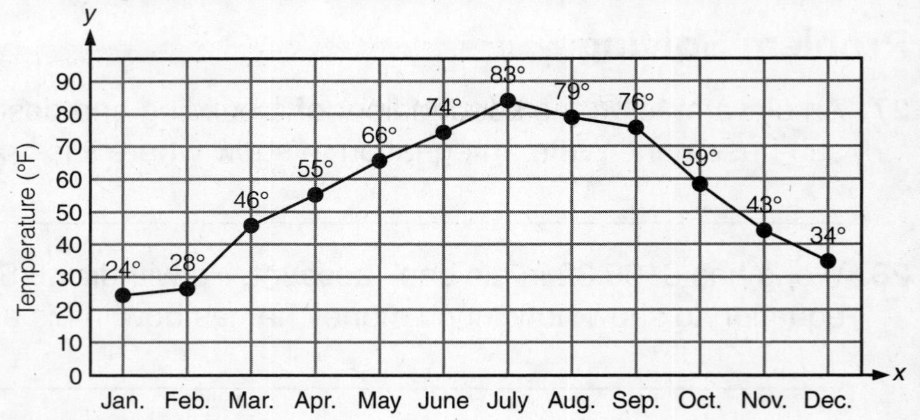

6. How much lower was the average daily high temperature in February than in August? ____________

Spiral Review

7. 3 m = ____________ mm

8. 125 cm = ____________ m

9. 4,900 g = ____________ kg

10. 83 L = ____________ mL

Name ____

Explore Addition and Subtraction Expressions

Write an expression for each situation.

1. There are c cars in a parking lot. Then 7 more cars drive into the lot. How many cars are in the lot now?

   ____

2. Diane has f oranges. She gives 3 of her oranges to Ben. How many oranges does she have now?

   ____

3. Julio talks to his grandmother for t minutes. He talks to her for another $5\frac{1}{2}$ minutes. How long does he talk to her in all?

   ____

4. Alison's family orders a pizza. After everyone takes a slice, there are p slices left. Alison takes $1\frac{1}{2}$ more slices. How many slices are left?

   ____

Evaluate each expression for the value given.

5. $b + 15$ for $b = 6$ ________
6. $w - 9$ for $w = 13$ ________
7. $14 + j$ for $j = 18.3$ ________
8. $19.7 - s$ for $s = 12.1$ ________

Solve.

9. Kim has run r meters. She runs 13 more meters. If $r = 87$, how many meters does Kim run in all? ____________

10. Gary has been studying for s minutes. He studies for $21\frac{1}{2}$ more minutes. If $s = 32$ minutes, how long does Gary study in all?

    ____

Spiral Review

Find the GCF for each pair of numbers.

11. 9 and 24 ________
12. 20 and 36 ________
13. 15 and 18 ________
14. 30 and 36 ________

Name ______________________________

Explore Multiplication and Division Expressions

Write an expression for each situation.

1. Roberto eats t oranges each week. How many oranges will he eat in 5 weeks?

____

2. Leah has d ounces of juice in a glass. She drinks $\frac{1}{3}$ of the juice. How many ounces of juice does she drink?

____

Evaluate each expression for the value given.

3. $4g$ for $g = 29$ ____

4. $\frac{z}{6}$ for $z = 96$ ____

5. $\frac{k}{5}$ for $k = 225$ ____

6. $\frac{3}{4}a$ for $a = 28$ ____

7. $7f$ for $f = 9$ ____

8. $5.2n$ for $n = 7$ ____

9. $\frac{3}{8}h$ for $h = 48$ ____

10. $\frac{u}{4}$ for $u = \frac{4}{9}$ ____

11. $\frac{2}{3}c$ for $c = \frac{6}{5}$ ____

12. $6.25x$ for $x = 7$ ____

13. $\frac{b}{8}$ for $b = 72$ ____

14. $0.2y$ for $y = 7.5$ ____

Solve.

15. Alicen has chosen a 250-page book for her book report. If she reads t pages a day, how long will it take her to read the book? ______

16. Angel is helping pack grapefruits for the historical society fund-raiser. If Angel can pack n grapefruits in each box, how many grapefruits will he pack into 8 boxes? ______

Spiral Review

Add or subtract.

17. $4 - 2\frac{1}{3} =$ ____

18. $3\frac{2}{5} + 4\frac{4}{5} =$ ____

19. $2\frac{1}{4} + \frac{2}{3} =$ ____

20. $1\frac{5}{6} - \frac{1}{2} =$ ____

Name ____

Order of Operations

Simplify. Use the order of operations.

1. $8 \times 3 - (9 + 4) =$ ________
2. $18 \div 2 - 3 \times 2 =$ ________
3. $(5.2 + 7) \times 4 =$ ________
4. $67 - 7 \times 5 + 24 \div 3 =$ ________
5. $(26 + 14) \div 4 + 7 =$ ________
6. $5^2 + 12 \div 2 =$ ________

Evaluate the expression for the value given.

7. $27 + h - 18$ for $h = 14$ ________
8. $55 - k \div 6$ for $k = 30$ ________
9. $401 + 18 \times z$ for $z = 10$ ________
10. $32 + 55 \times b \div 11$ for $b = 2$ ________

Place parentheses to make the sentence true.

11. $10 \times 8 - 5 + 7 = 37$ ____________________
12. $55 - 15 \div 3 + 2 = 8$ ____________________

Problem Solving

13. Antoine orders 4 boxes of daffodil bulbs and 7 boxes of tulip bulbs to plant in his garden. Each box of daffodil bulbs contains 12 bulbs, and each box of tulip bulbs contains 10 bulbs. Write an expression to find out how many more tulips than daffodils Antoine ordered. Then evaluate the expression.

____

14. Elena goes to the bank and gets 17 rolls of quarters and 12 rolls of dimes. If each roll of quarters contains 40 coins, and each roll of dimes contains 50 coins, write an expression for the total number of coins that Elena gets. Then evaluate the expression.

____

Spiral Review

Complete the equivalent fraction.

15. $\frac{3}{4} = \frac{\quad}{12}$
16. $\frac{14}{20} = \frac{7}{\quad}$
17. $\frac{1}{2} = \frac{\quad}{84}$
18. $\frac{\quad}{7} = \frac{30}{35}$

Name ______________________________

Functions

Write an equation to describe the situation. Tell what each variable represents.

1. Victor is sending an express package to his father through a delivery service. The service charges $2.50 per pound, plus a $10.00 fee for express delivery.

____

____

Complete the table. Write an equation to describe the situation.

2. Calculating Taxi Fare

Cost to hire a taxi: $1.50

Number of miles	1	2	3	4	5
Total Fare	$3.50	$5.50	$7.50		

____

Problem Solving

3. Gina is ordering CDs over the Internet. There is a $6 shipping charge, and each CD costs $12. Gina orders 4 CDs. Write an equation to describe the situation. Tell what the variables represent. Use your equation to solve the problem.

____

____

4. Juan is going to ride his bicycle 12 blocks to his grandmother's house. It takes him 5 minutes to get ready and leave his house. It then takes him $\frac{1}{2}$ a minute to ride each block. Write an equation to describe the situation. Tell what the variables represent. Use your equation to solve the problem.

____

____

Spiral Review

5. 330 min = ________ h 6. 10,560 ft = ________ mi 7. 56 oz = ________ lb

Name ______________________________

Graphing a Function

Write the coordinates for each point.

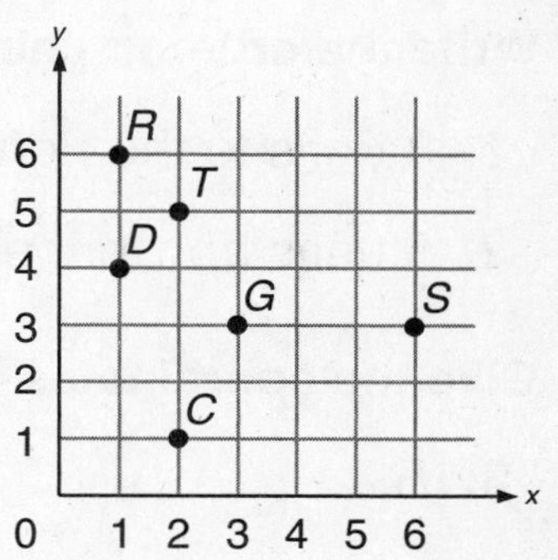

1. *C* ______ **2.** *D* ______ **3.** *G* ______

Name the point for each ordered pair.

4. (1, 6) ______ **5.** (6, 3) ______ **6.** (2, 5) ______

Complete the table. Then graph the function.

7. $m = 2n + 1$

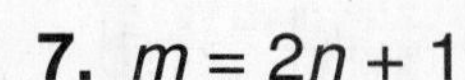

n	*m*
0	1
1	3
2	
3	

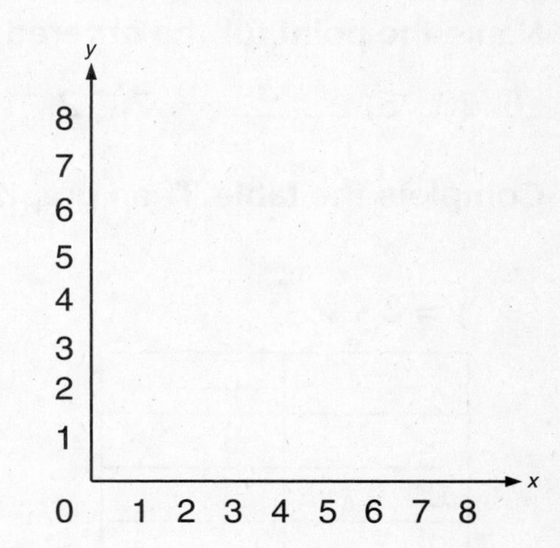

Problem Solving

8. Debra makes $3 an hour baby-sitting. She also gets a $2 tip each time she baby-sits. Write an expression that describes the relationship between the number of hours that she baby-sits, *h,* and the amount of money she earns, *m.* ______________

9. The taxi to the airport charges $1.35 per mile plus the cost of tolls. If the parkway toll is $0.25 and the bridge toll is $3.50, what is the cost of a trip to the airport? Write your answer as an expression describing the relationship between miles and total cost. ______________

Spiral Review

10. $\frac{2}{7} \times \frac{2}{5} =$ ___ **11.** $\frac{4}{9} \div \frac{1}{6} =$ ___ **12.** $\frac{5}{8} \times \frac{7}{10} =$ ___

Name ______________________________

Graph in Four Quadrants and Solve Problems Using Graphs

Write the ordered pair for each point described.

1. 3 units to the right of the origin; 2 units below the origin ________

2. 5 units above the origin; 1 unit to the left of the origin ________

Give the coordinates of the point.

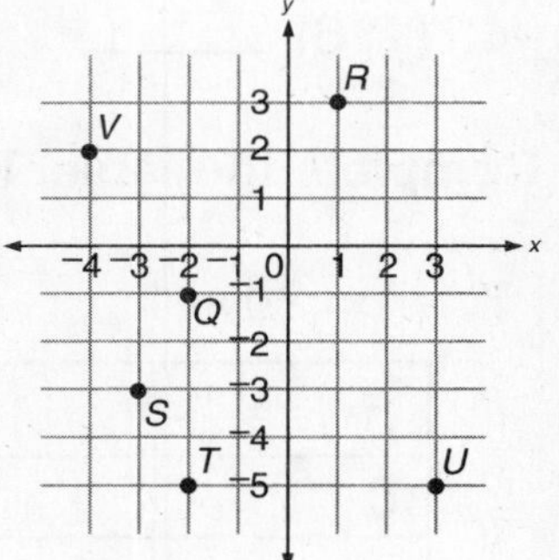

3. *R* ________ **4.** *T* ________ **5.** *V* ________

Name the point for the ordered pair.

6. (3, $^{-}5$) ________ **7.** ($^{-}2$, $^{-}1$) ________ **8.** ($^{-}3$, 3) ________

Complete the table. Then graph the function.

9. $y = 3 - x$

x	*y*
$^{-}2$	5
0	
2	
4	

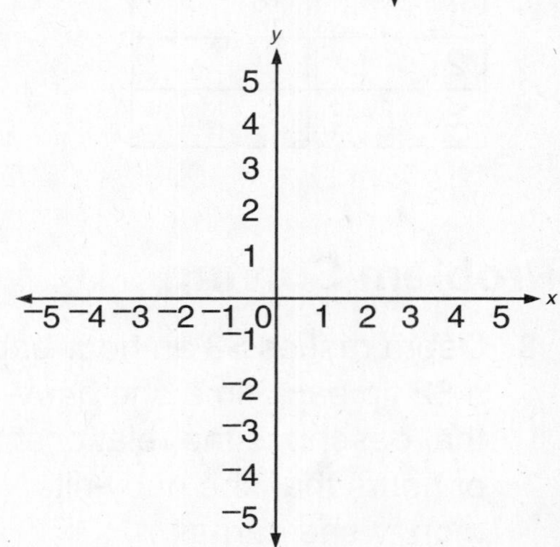

Problem Solving

10. Linda is graphing the equation $y = x + 4$. Will the coordinates (1, 5) show up on her graph?

____

11. John's graph of a straight line function begins at point ($^{-}3$, $^{-}3$) and ends at point (4, 2). Through how many quadrants of the graph does his line run?

____

Spiral Review

Find the LCD.

12. 4 and 5 ________ **13.** 6 and 8 ________ **14.** 4 and 15 ________

Name ______________________________

Problem Solving: Reading for Math

Use Graphs to Identify Relationships

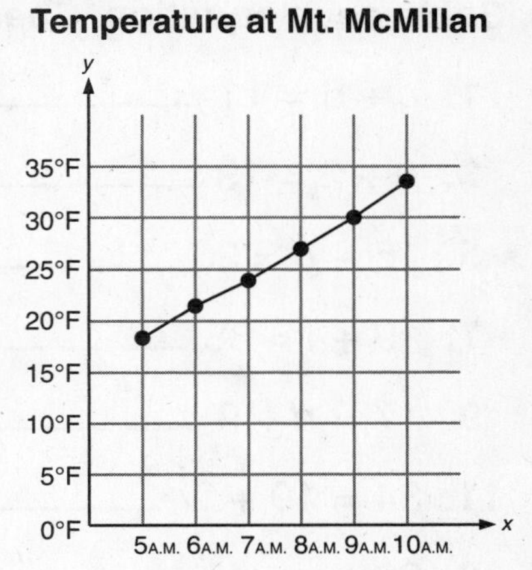

1. The graph shows the air temperature at the top of Mount McMillan from 5 A.M. to 10 A.M. on March 13. How would you describe the relationship between time and air temperature as shown in this graph?

____

2. From 10 A.M. on, the air temperature stayed at 34°F. How would this be shown on a graph?

____

Use data from the graph for problems 3–5.

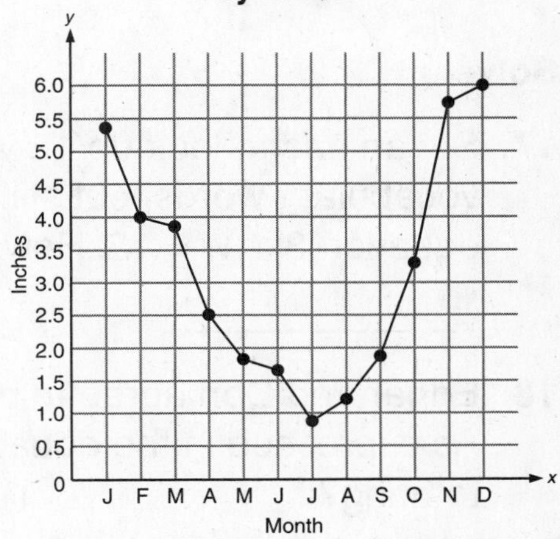

source: *World Almanac and Book of Facts 2000*

3. The graph at the right shows the average monthly precipitation in Seattle, Washington, for each month of the year. What change occurs in the average precipitation between January and February? between February and March? between March and April?

____

____

4. How would you describe the relationship between month and precipitation from January to July?

____

5. What is the difference in precipitation between the wettest month and the driest month?

____

Spiral Review

6. $1.4 \times 2.3 =$ ______ 7. $9.54 \div 0.6 =$ ______ 8. $191.4 \div 3.3 =$ ______

Name ____

Explore Addition Equations

Solve each equation. Check your answer.

1. $c + 5 = 11$ ________
2. $j + 4 = 22$ ________
3. $p + 2 = 19$ ________
4. $n + 9 = 34$ ________
5. $15 + g = 23$ ________
6. $23 + b = 36$ ________
7. $19 + t = 38$ ________
8. $12 + f = 26$ ________
9. $17 = d + 9$ ________
10. $28 = e + 12$ ________
11. $34 = 19 + w$ ________
12. $41 = 34 + q$ ________
13. $29 + r = 46$ ________
14. $z + 12 = 23$ ________
15. $22 = h + 9$ ________
16. $51 = a + 18$ ________

Solve.

17. Susan has to look up 12 vocabulary words for school. She has looked up 3 vocabulary words, but still has w words to look up, as expressed in the equation $3 + w = 12$. How many words does she still have to look up?

____

18. Edgar and Don agree to bring 23 cans of fruit juice to a picnic. Don brings 10. As expressed in the equation $23 = 10 + e$, how many cans does Edgar have to bring?

____

19. The baby has spilled Julia's crayons all over the floor, and Julia must pick them up and put them back into the box. The box contains 48 crayons, and Julia has already picked up 22. As expressed in the equation $22 + c = 48$, how many more crayons must Julia pick up?

____

Spiral Review

20. $120 \times \frac{2}{3} =$ ________
21. $600 \div \frac{2}{5} =$ ________
22. $360 \div \frac{3}{4} =$ ________

Name ____________________

Addition and Subtraction Equations

Solve each equation. Check your answer.

1. $h + 9 = 21$ __________
2. $k + 34 = 71$ __________
3. $u - 8 = 72$ __________
4. $a - 7.3 = 15.6$ __________
5. $131 + m = 147$ __________
6. $f - 19 = 92$ __________
7. $y + \frac{3}{5} = 35\frac{1}{5}$ __________
8. $326.7 - s = 297.4$ __________
9. $m + 5 = {}^{-}1$ __________
10. $u + ({}^{-}3) = 7$ __________

Without solving each equation, tell whether the solution is greater than 44, less than 44, or equal to 44.

11. $r + 10 = 44$ __________
12. $x - 56 = 44$ __________

Problem Solving

13. The gas tank in Maria's mother's car can hold 14 gallons of gas. If her mother fills the tank by putting in 9 gallons of gas, how much gas was already in the tank? Write an equation and solve it.

____

14. The road from Chicago to Minneapolis goes through Madison, Wisconsin. It is 409 miles from Chicago to Minneapolis, and 146 miles from Chicago to Madison. How far is it from Madison to Minneapolis? Write an equation and solve it.

____

Spiral Review

Identify each property.

15. (2,903 + 4,255) + 860 = 2,903 + (4,255 + 860)

____

16. $92{,}754 \times 0 = 0$

____

Name ____________________

Multiplication and Division Equations

For each equation, decide if you should multiply or divide both sides to solve.

1. $7y = 21$ __________
2. $f \div 2 = 19$ __________
3. $15x = 125$ __________
4. $\frac{s}{3} = 51$ __________

Solve each equation. Check your answer.

5. $u \times 6 = 618$ __________
6. $\frac{y}{7} = 32$ __________
7. $t \times 4.2 = 21$ __________
8. $\frac{p}{3.5} = 3$ __________
9. $12n = 1{,}440$ __________
10. $125 \div t = 25$ __________
11. $w \times 8 = 680$ __________
12. $336 \div g = 96$ __________

Choose the equation for which the value of *n* is a solution. Circle your answer.

13. $n = 5$

 A. $n + 2 = 9$ B. $2n = 10$ C. $\frac{n}{2} = 3$

14. $n = 3.2$

 A. $7.7 - n = 4.5$ B. $10n = 35$ C. $n \div 0.5 = 6$

Problem Solving

15. At the grocery store, 16 oz of apples costs \$2.40, 24 oz costs \$3.60, and 32 oz costs \$4.80. Write an equation that relates the price of the apples to their weight. __________

16. According to the equation from problem 13, how much will 132 oz of apples cost? __________

Spiral Review

Describe each number as prime or composite.

17. 17 __________
18. 49 __________
19. 230 __________

Name ____

Problem Solving: Strategy

Make a Graph

Use the graph to solve.

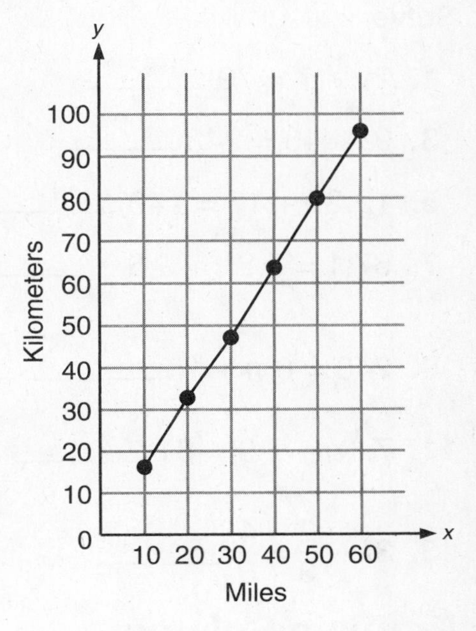

1. It is 44 miles from San Francisco to San Jose. About what is this distance in kilometers?

____

2. It is 48 kilometers from Dallas to Fort Worth. About what is this distance in miles?

____

3. Steve travels 50 miles to visit his grandmother, and decides to check the graph to see how far he has traveled in kilometers. He estimates that he has traveled about 30 kilometers. Is he reading the graph correctly?

____

4. Pam drives 15 miles to the beach. Going home, she takes a different route, and drives 18 miles. About how far does she drive in kilometers?

____

Mixed Strategy Review

5. Electra owns 4 skirts. If she wants to wear a different outfit every day for a full month, how many blouses must she have? __________

6. Harry would like to surprise his mother with a bouquet of flowers for her birthday. The flowers cost $38 plus a 6% sales tax. The delivery charge is $5. Harry has one $20 bill, one $10 bill, and eight $1 bills. How much more money does Harry need? __________

Spiral Review

Name the place-value of the 6 in each problem.

7. 41.68 ________________ **8.** 62,993.3 ________________

9. 2,910.96 ________________ **10.** 56.207 ________________

Name ________________________________

Two-Step Equations

Solve.

1. $2y + 7 = 79$ ______

2. $4u - 9 = 91$ ______

3. $9t + 45 = 243$ ______

4. $551 = 6f - 19$ ______

5. $12.3e + 17 = 140$ ______

6. $11r - 19 = 102$ ______

7. $67.1 = 4.8h + 9.5$ ______

8. $\frac{p}{4} + 4 = 19$ ______

9. $240 = 6m + 18$ ______

10. $\frac{38}{s} - 7 = 12$ ______

11. $7.1q + 2.9 = 81$ ______

12. $\frac{w}{3} - \frac{5}{6} = \frac{1}{2}$ ______

13. $3k - \frac{2}{3} = \frac{1}{3}$ ______

14. $4v + 7 = {}^{-}9$ ______

Problem Solving

15. Candice has 23 plums. She bought 8 of these at the grocery store and was given the rest by her neighbor. Her neighbor gave Candice exactly $\frac{1}{4}$ of the plums he got from his plum tree. How many plums did her neighbor get from his tree? ______

16. The fifth grade at Pierce Elementary School is going on a field trip to the art museum. There are 119 students and teachers going in all. Of these, 7 students are being driven to the museum by their parents. Everyone else will go in 4 school buses. How many people will be on each bus if each bus has the same number of people on it? ______

Spiral Review

Compare. Write >, <, or =.

17. $\frac{1}{8}$ ______ $\frac{2}{9}$

18. $\frac{3}{4}$ ______ 0.7

19. 4.6 ______ $4\frac{3}{5}$

Name ______________________________

Basic Geometric Ideas

Identify the figure. Then name it using symbols.

1. F G (ray) ______________ **2.** M N (segment) ______________

Use data from the diagram for problems 3–4.

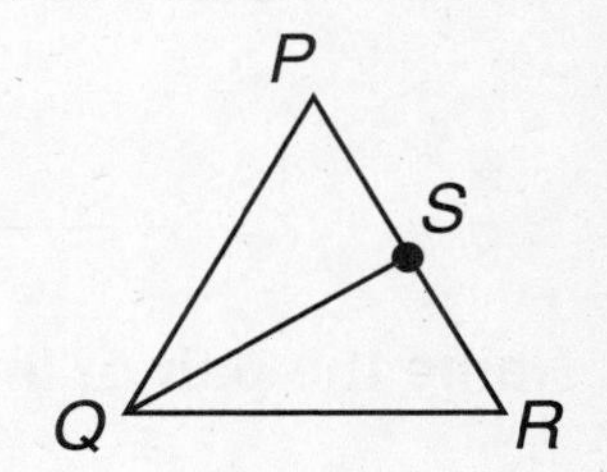

3. Name all the points. ______________

4. Name a line segment.

____

____

Tell which line segments are congruent. Use a ruler to measure.

5. H I, J K ________ **6.** W X, Y Z ________

Write if the figure is a polygon. If not, explain why.

7.

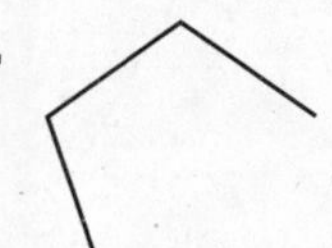

____

____

8.

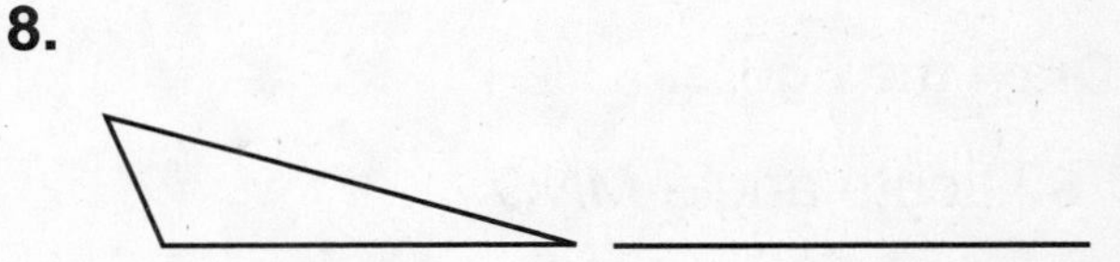

____

Problem Solving

9. Dolores notices that a window in her father's car is made up of four line segments that meet to form four vertices. Identify the figure formed by the window.

10. The Pentagon Building is the headquarters of the Department of Defense. Based on its name, how many sides should it have? How many vertices?

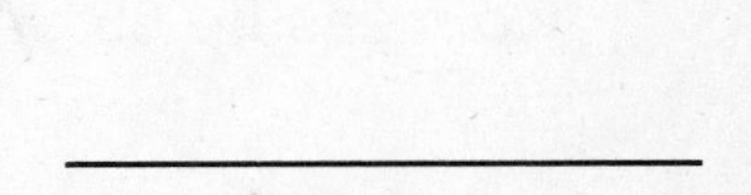

Spiral Review

Evaluate each expression for the value given.

11. $7u$ for $u = 23$ ______________ **12.** $\frac{t}{3}$ for $t = 684$ ______________

Name ____

11•2 Measure and Classify Angles

Use a protractor to measure each angle. Classify the angle as acute, right, or obtuse.

1. ________________

2. ________________

3. ________________

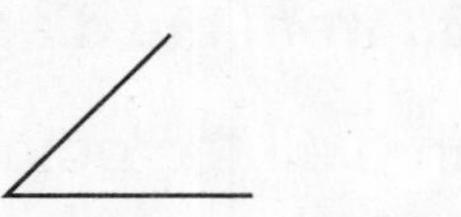

Name the pair of lines as intersecting, parallel, or perpendicular.

4. ____________________

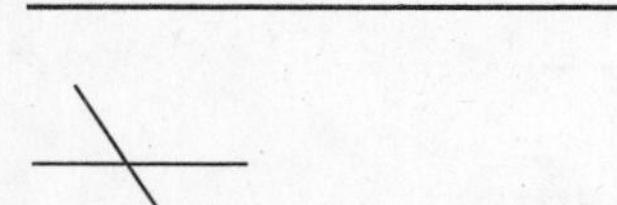

5. ____________________

6. ____________________

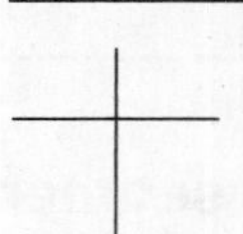

Find the measure of the following angles in the diagram. Classify each angle as acute, right, or obtuse.

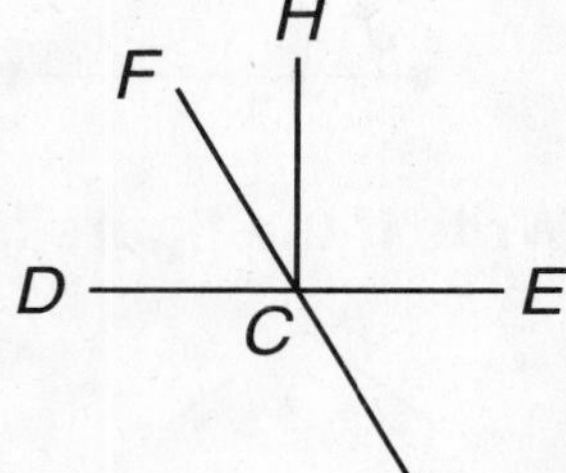

7. *DCF* ____________________

8. *ECF* ____________________

Draw the figure.

9. acute angle *MNO*

10. right angle *BCD*

Problem Solving

11. A 36° angle is placed together with a 45° angle. Is the new angle they form acute, right, or obtuse?

12. Ellis measures the four angles formed by the intersection of *JK* and *LM.* Two angles measure 125° and the other two measure 55°. Are the lines perpendicular?

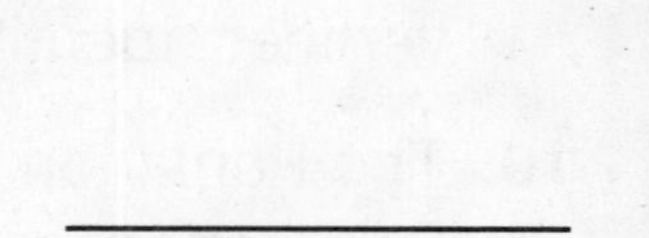

Spiral Review

13. $\frac{4}{5} = \frac{}{20}$

14. $\frac{}{8} = \frac{12}{32}$

15. $\frac{2}{7} = \frac{26}{}$

16. $\frac{5}{16} = \frac{}{80}$

Name ______________________________

Triangles

Classify each triangle as equilateral, isosceles, or scalene and right, acute, or obtuse.

1. ______________ **2.** ______________ **3.** ______________

 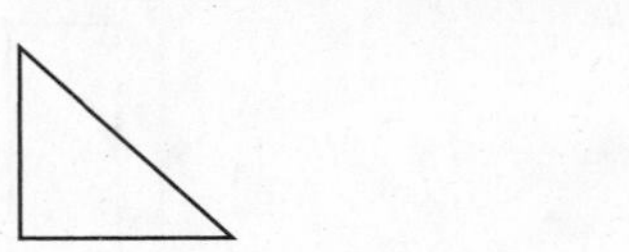 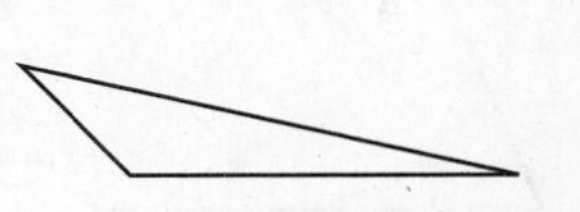

4. ______________ **5.** ______________ **6.** ______________

 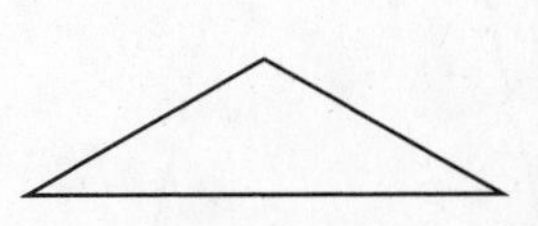

Use a ruler and protractor to draw these triangles and then classify them.

7. All angles less than 90 degrees, two sides congruent

____

8. One angle greater than 90 degrees, no congruent sides

____

Problem Solving

9. Can an equilateral triangle be anything other than acute? Why or why not?

____

____

10. Tim measures two sides of a triangle and finds that they are congruent. Can he classify the triangle as isosceles? Why or why not?

____

Spiral Review

11. $1{,}477 \times 24 =$ ______________ **12.** $45.48 \times 1.5 =$ ______________

Name ____

Quadrilaterals

Name the quadrilateral in as many ways as you can. Write the sum of the measure of the angles for each one.

1. ______________________________

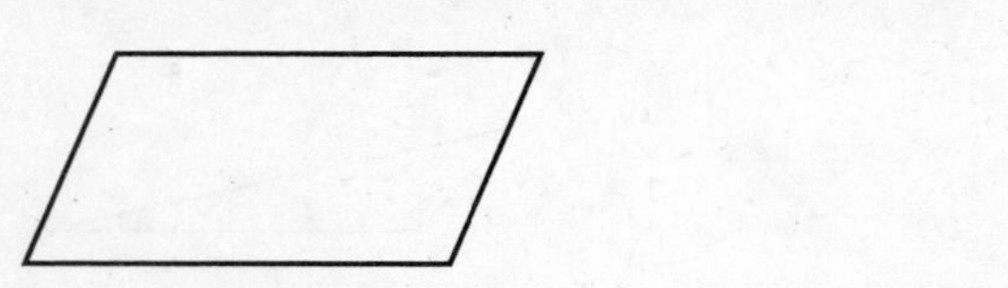

2. ______________________________

3. ______________________________

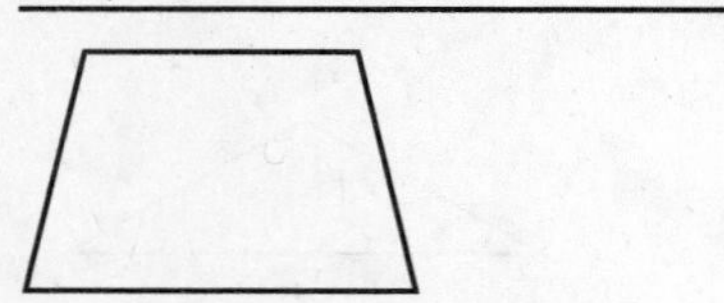

4. ______________________________

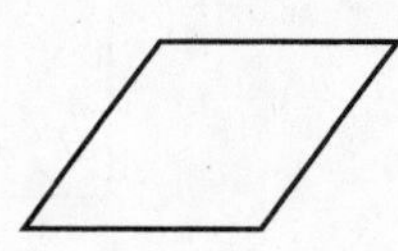

Write true or false.

5. All parallelograms are squares.

____

6. All squares are parallelograms.

____

7. The sum of the angles of a trapezoid is less than 360°.

____

8. All squares are rhombuses.

____

Problem Solving

9. Bob notices a quadrilateral in a mural. The quadrilateral has two angles of 75° and two angles of 105°. It also has two pairs of congruent sides that are parallel with one another. What is the name of the quadrilateral?

____

10. Vera is measuring the angles of a trapezoid. She measures the first three to be 100°, 65°, and 75°. What is the measure of the fourth angle?

____

Spiral Review

Find the LCD.

11. 4 and 10 ________

12. 6 and 10 ________

13. 8 and 9 ________

Name ______________________________

Problem Solving: Reading for Math

Draw a Diagram

Draw a diagram to solve each problem.

1. Jorge plays the triangle in a band. He orders a new triangle by mail. The triangle is equilateral, with sides of 10 inches. What is the smallest rectangular box the triangle can be shipped in?

____

2. Renata orders a poster that is 11 inches by 13 inches, and music that is 10 inches by 15 inches. In what size and shape box will her order be sent?

____

3. Tom orders a music stand. The stand is made up of a square easel that is 17 inches by 17 inches, and a 3-foot rod that attaches to the top of the easel. What is the smallest size rectangular box that the stand can be shipped in?

____

Use data from the diagrams for problems 4–6.

4 ft
10 in.

14 in.
14 in.

1 ft
10 in. 10 in.
2 ft

4. The figures above show choices available at Board World, a lumber store. Can a trapezoid be shipped in a box that measures 2 feet by 2 feet?

____

5. Hal orders a square. Which of the following boxes could possibly contain Hal's shipment? 3 feet by 1 foot; 2 feet by 15 inches; 10 inches by 20 inches

____

6. What is the smallest rectangular box that an order for a rectangle and a square could be shipped in?

____

Spiral Review

7. $\frac{4}{5} \times \frac{6}{7} =$ ____

8. $\frac{9}{4} \div \frac{3}{8} =$ ____

9. $\frac{11}{2} \times \frac{5}{3} =$ ____

Name ___________________________________

11•6 Congruence and Similarity

Tell whether the figures are congruent, similar, or neither.

1. 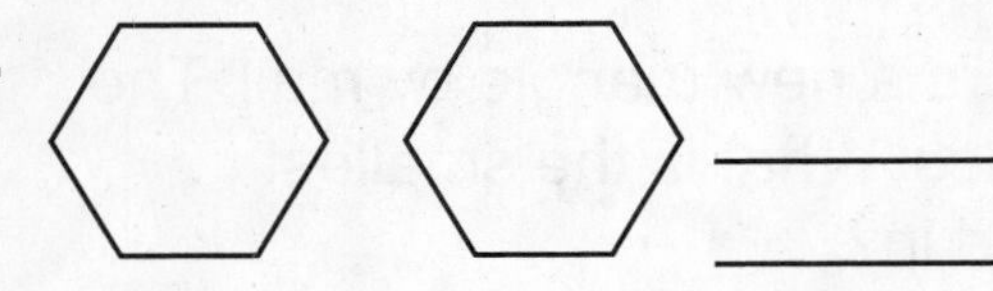 ____________

2. 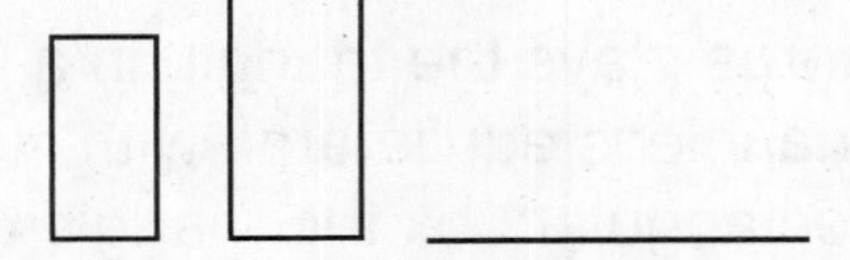 ____________

3. 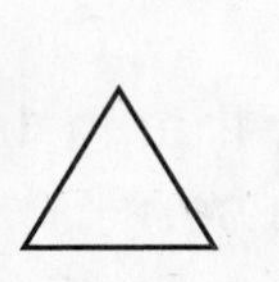  ____________

4. 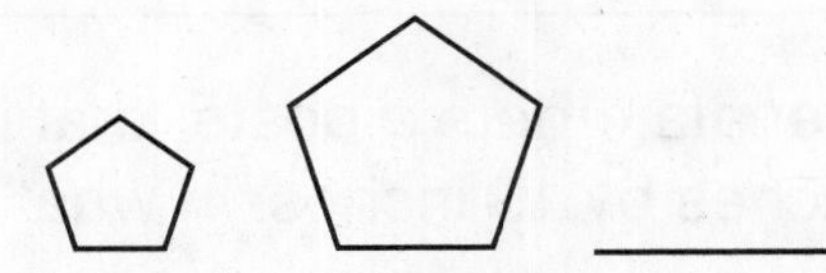 ____________

Find the measure of the missing angle in each pair of similar figures.

5. 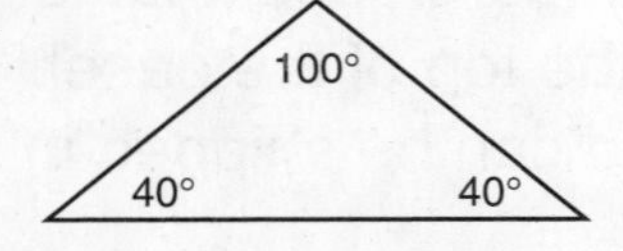

____

6. 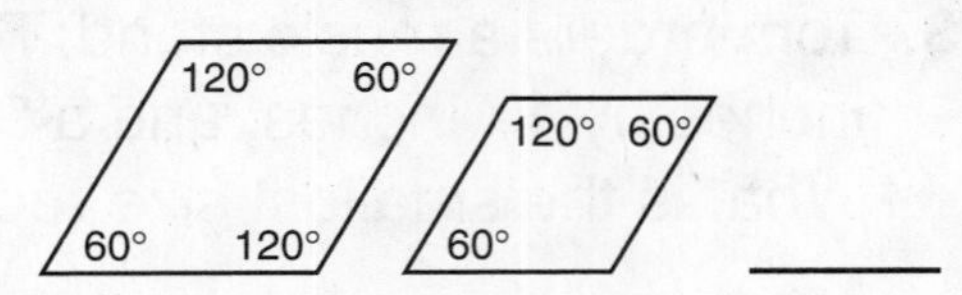

____

Identify the corresponding side or angle.

7. $\overline{BC}$ ____________

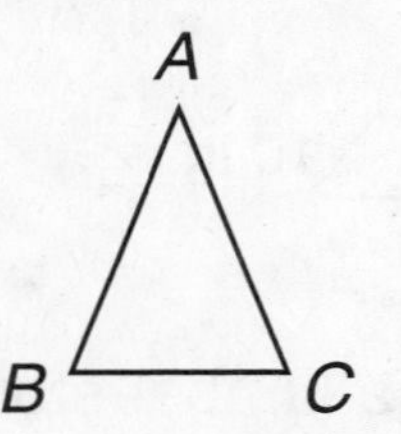

8. $\overline{TU}$, ____________

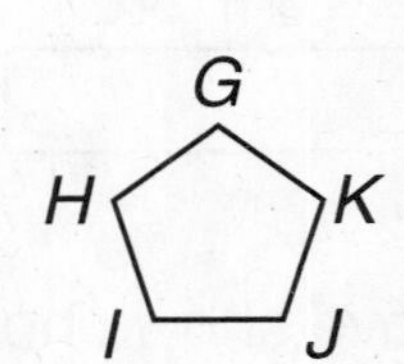

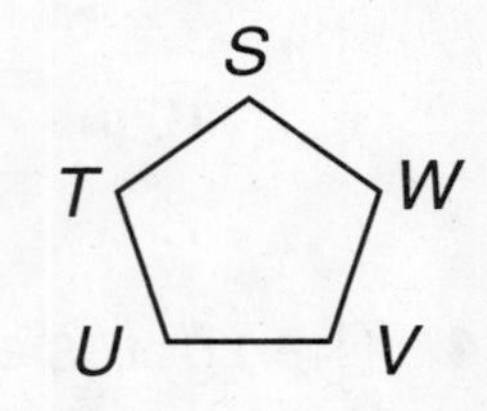

Problem Solving

9. There are two large windows in Ms. Fehler's classroom. Both measure 6 feet wide by 4 feet tall. Are the two windows congruent? ____________

Spiral Review

Solve.

10. $n + 17 = 39$ ________

11. $u - 48 = 101$ ________

Name________________________________

11•7 Transformations

Write whether a translation, reflection, or rotation was made.

1. 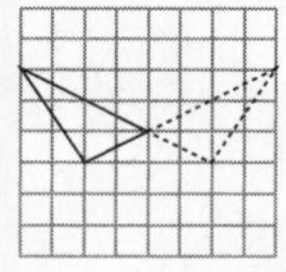

____

2. 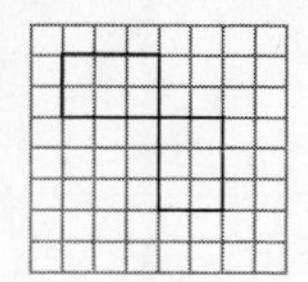

____

3. 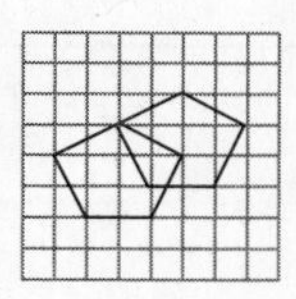

____

Choose the figure that is made by the transformation described.

4. Rotation

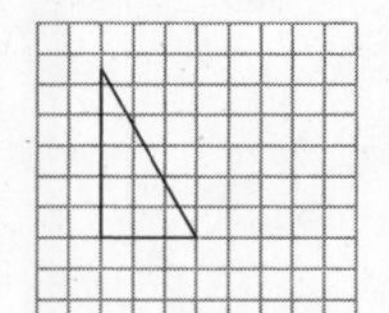

A. 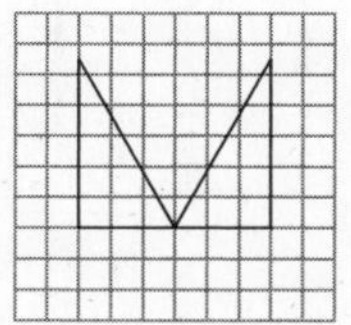

B. 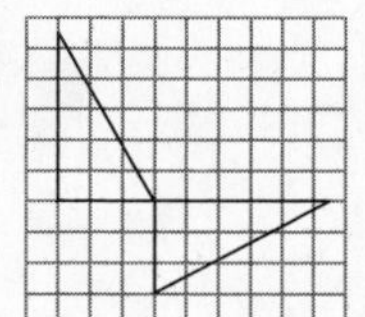

C. 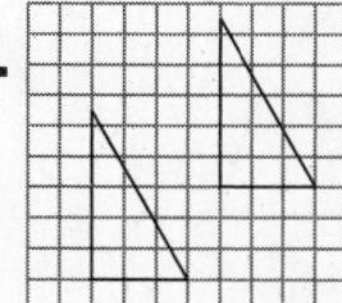

5. Translation

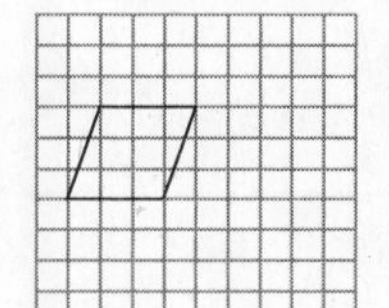

A. 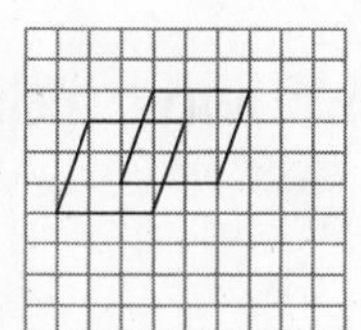

B. 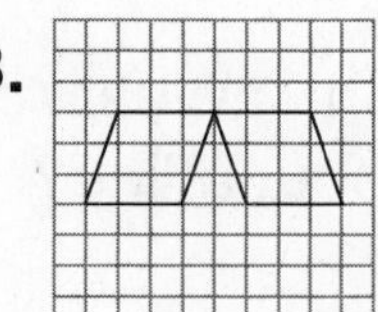

C. 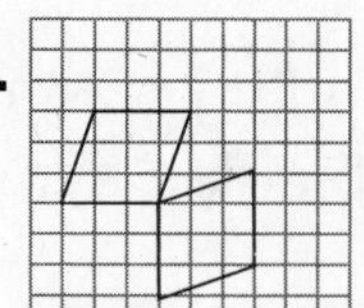

Problem Solving

6. The flag of Iceland is shown. If the flag is reflected, will it look the same?

____

7. In an abstract painting, two congruent squares sit next to one another as shown. What are two possible ways to describe the transformation that these squares show?

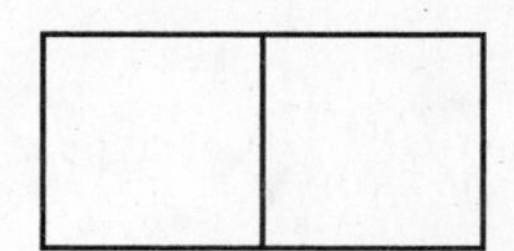

____

Spiral Review

Simplify. Use the order of operations.

8. $51 + 17 \times 8 =$ ____________

9. $228 \div 2^2 - 15 =$ ____________

Name ____________________

11•8 Symmetry

Tell which figures are symmetric about a line.

1. 
____

2.
____

3. 
____

Draw all the lines of symmetry for each figure.

4. 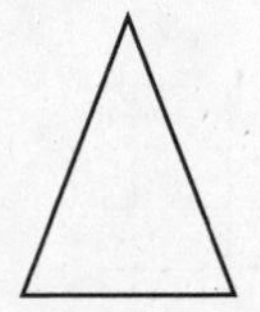

5. 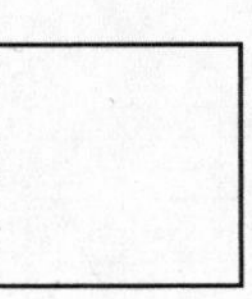

Problem Solving

6. The flag of Mali, a nation in West Africa, is shown below. Is the flag symmetric about a line? ______________

7. The shape of the nation of Mali is shown below. Is the shape of the nation symmetric about a line? ______________

Spiral Review

Solve.

8. $5c + 17 = 192$ ________

9. $\frac{x}{8} - 12 = 56$ ________

Name ______________________________

11•9 Problem Solving: Strategy

Find a Pattern

Solve.

1. Jody is drawing the pattern shown above. Use a ruler to measure the dimensions of the rectangles in the pattern. If Jody continues the pattern in the same way, what will be the dimensions of the sixth rectangle she draws?

____

2. Instead of continuing, Jody begins to count back down, so that the fifth rectangle is 1 × 3. What will the dimensions of the seventh rectangle be?

____

3. Luke draws a pattern that also starts with a 1 cm × 1 cm square. The second figure in his pattern is a 4 cm × 4 cm square, and the third figure is a 7 cm × 7 cm square. What will the dimensions of the seventh figure be?

____

4. Sarah is making a pyramid out of blocks. She begins with a base of 15 blocks × 15 blocks. She then gives the second level dimensions of 13 blocks × 13 blocks, and the third level dimensions of 11 blocks × 11 blocks. How many blocks will it take to construct the fifth level?

____

5. Sarah decides to make a pyramid that follows the same pattern as the first one but is one level taller. How many levels will the second pyramid have?

____

6. What will the dimensions of the base of the second pyramid be?

____

Spiral Review

Find the GCF.

7. 15 and 40 ____

8. 16 and 56 ____

9. 148 and 100 ____

10. 72 and 108 ____

Name ______________________________

Circles

Identify the parts of circle *E*.

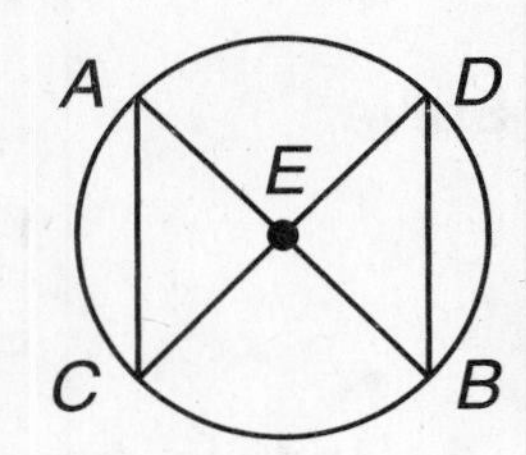

1. Center ____________
2. Chords ____________
3. Radii ____________
4. Diameters ____________

Solve. Use circle *E*.

5. If $CD = 7$ feet, how long is AB? ____________
6. A new chord BC is drawn on the circle. Is this chord a diameter? ____________

Draw circles with the given measurements.

7. diameter = 2 inches

8. radius = 2 inches

Problem Solving

9. Tamara has a circular wall clock. She imagines the following line segment running through the clock. What part of a circle is she imagining?

____

10. A basketball hoop has a radius of 9 inches. What is its diameter?

____

Spiral Review

11. $3\frac{1}{2} - 1\frac{4}{5} =$

12. $2\frac{2}{3} + 4\frac{3}{4} =$

Name ____

11•11 Explore Tessellations

Tell whether the shapes tessellate. Show your work.

1. ______________________________

2. ______________________________

3. ______________________________

4. ______________________________

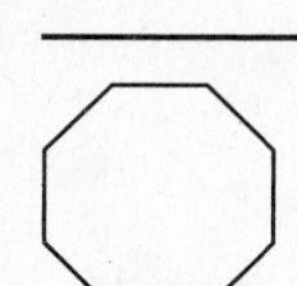

5. ______________________________

6. ______________________________

Solve.

7. A rectangular window in Hiram's house is made up of square panes as shown. Do the panes in the window tessellate?

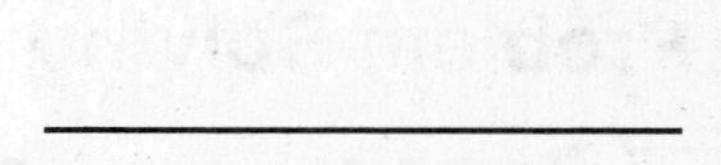

8. Elaine is making a pattern with oak leaves like the one shown here. Will the oak leaves in her pattern tessellate?

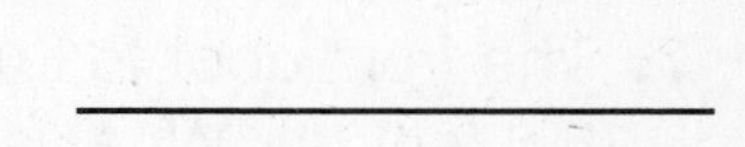

Spiral Review

Evaluate each expression for the value given.

9. $\frac{s}{3}$ for $s = {}^{-}12$ __________

10. $8k$ for $k = {}^{-}2$ __________

Name ______________________________

Perimeter of Polygons

Find the perimeter of each figure.

1. ______________

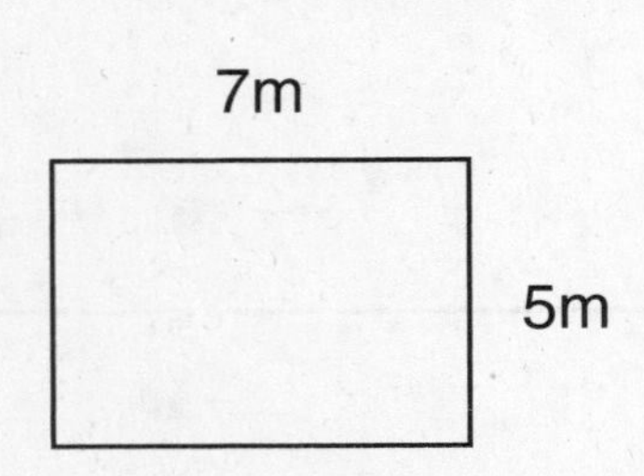

2. ______________

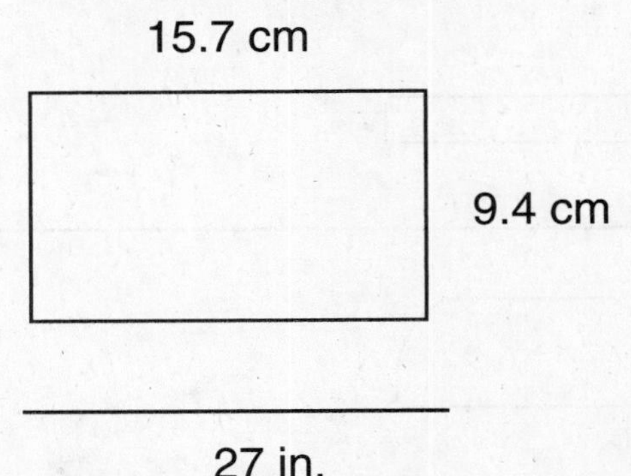

3. ______________

9 ft

9 ft

4. ______________

27 in.

17 in.

Problem Solving

5. Mona has a rectangular painting that measures 6 feet wide by 4 feet tall. What is the perimeter of the painting? ______________

6. A vacant lot measures 57 meters wide by 73 meters long. How many meters of fencing will be needed to fence in the entire lot? ______________

7. The front door to Lucy's house measures 3.1 feet wide by 6.8 feet tall. What is the perimeter of the door? ______________

Spiral Review

Solve.

8. $d + 13 = 77$ ________ **9.** $k - 28 = 125$ ________ **10.** $p + 7 = 1$ ________

Name ______________________________

Area of Rectangles

Find the area of each figure.

1. ______________

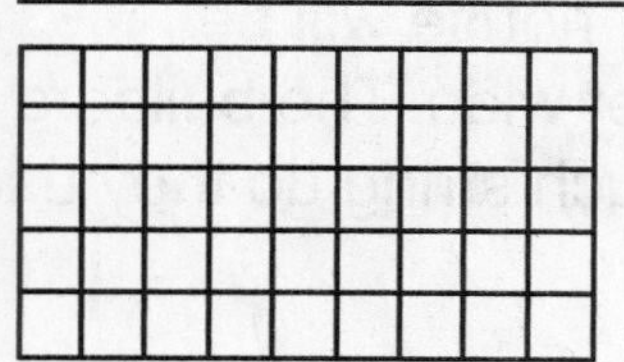

2. ______________

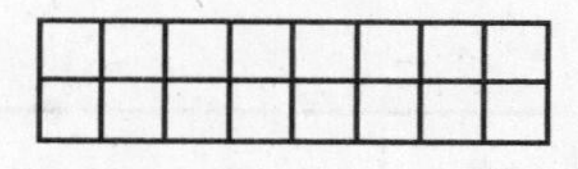

3. ______________

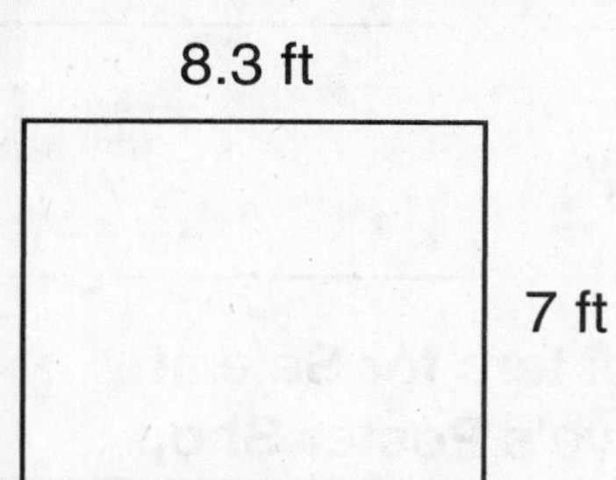

4. ______________

5.2 m

Problem Solving

5. Ms. Newton's classroom is a rectangle that measures 7 meters wide by 12 meters long. What is the area of the classroom? ______________

6. Anna notices that a sheet of notebook paper measures 11 inches tall and 8.5 inches wide. What is the area of the sheet of paper? ______________

7. Demetrius's backyard measures 60 feet long by 45 feet wide. What is the area of the backyard? ______________

Spiral Review

Compare. Write >, <, or =.

8. $\frac{4}{5}$ ______ 0.8

9. $\frac{4}{3}$ ______ $\frac{5}{4}$

10. $\frac{8}{7}$ ______ 1.2

Name ________________________________

Problem Solving: Reading for Math

Distinguish Between Perimeter and Area

State whether perimeter or area is needed. Then solve the problem.

1. A construction company is building a new house. The house will be rectangular, and it will measure 50 feet long by 32 feet wide. The builders are using string to mark where the house will be. How much string do they use?

   ____

2. The lot on which the house is being built measures 90 feet by 100 feet. How large is the lot? ________________

3. The owner of the house wants to have at least 5,000 ft^2 of space for a lawn. If the house is built as planned, will there be enough lawn? ________________

Use data from the table for problems 4–7.

Posters for Sale at Dave's Poster Shop

Poster	**Dimensions**
Basketball Players	40 in. × 30 in.
Grizzly Bear	2.5 ft × 2 ft
Snowy Mountains	5 ft × 4 ft
Skyscraper	10 in. × 50 in.

4. Scott wants to buy the Snowy Mountains poster but is not sure how much space it will take up on the wall in his room. How can he determine the space? How much space will that be?

   ____

5. Scott buys the Basketball Players poster and has it framed. How can he determine the total length of framing needed? What is that length?

   ____

6. Dina has a square space totaling 9 ft^2 for a poster. Will the Grizzly Bear poster fit in the space? ________________

7. Tyra buys the Skyscraper poster. She places the poster on a piece of cardboard and traces its outline. What is the length of the line she traces? ________________

Spiral Review

Solve.

8. $4x = 164$ ____________ 9. $\frac{y}{4} = 164$ ____________ 10. $\frac{333}{z} = 9$ ____________

Name ______________________________

Explore Area of Parallelograms

Find the area of each figure.

1. ____________

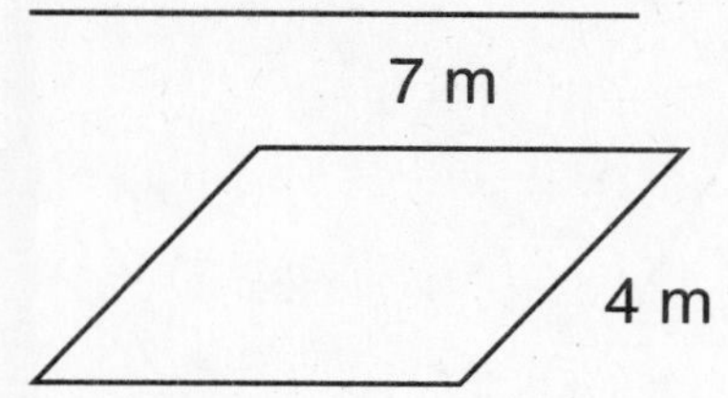

2. ____________

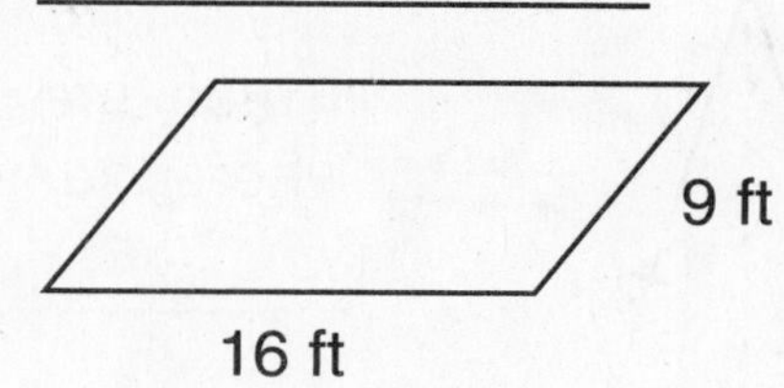

3. ____________

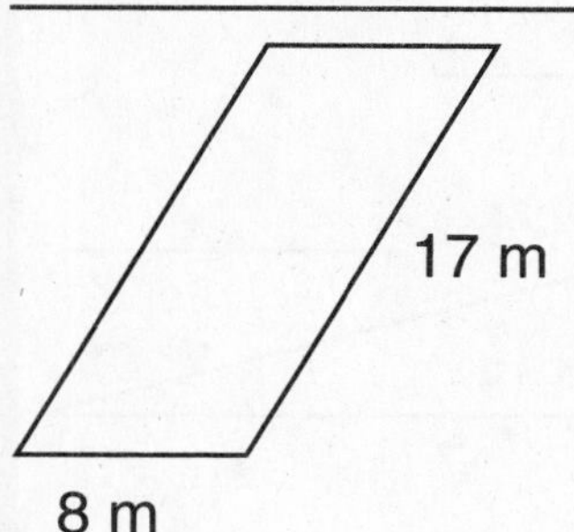

4. ____________

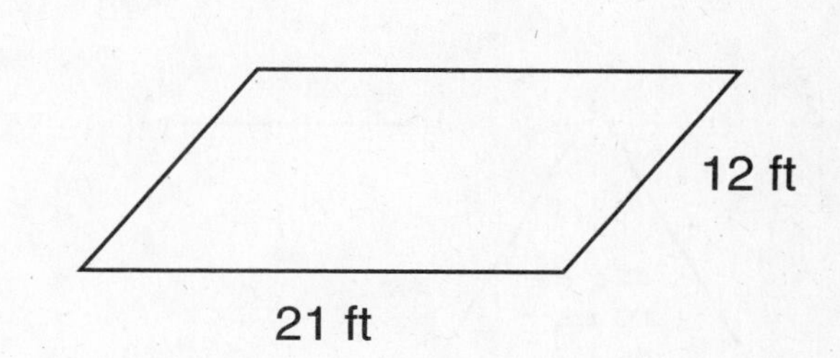

5. ____________

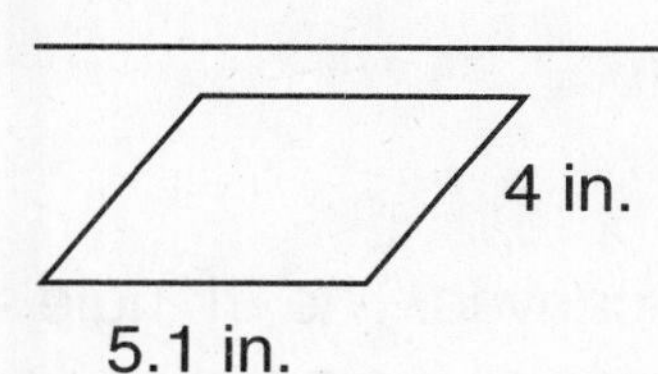

6. ____________

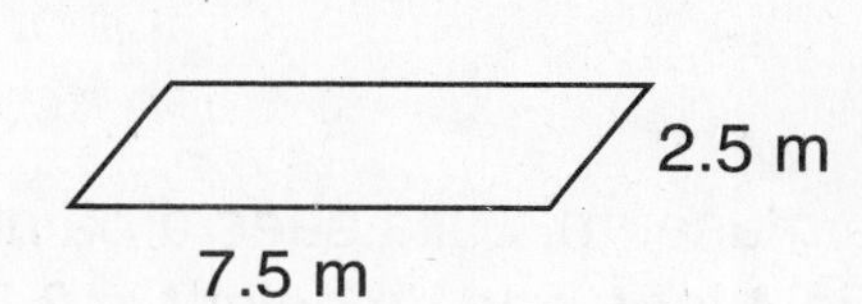

Solve.

7. While riding the bus, José notices that one of the windows is a parallelogram. If the base of the parallelogram is 14 inches and its height is 9 inches, what is its area? ____________

8. The lot on which Kelly's house is built has the shape of a parallelogram. The base of the lot is 35 meters, and its height is 20 meters. What is the area of the lot? ____________

Spiral Review

9. $7c - 29 = 209$ ____________

10. $\frac{p}{4} + 8 = 44$ ____________

Name ___

Explore Area of Triangles

Find the area of each triangle.

1. ___

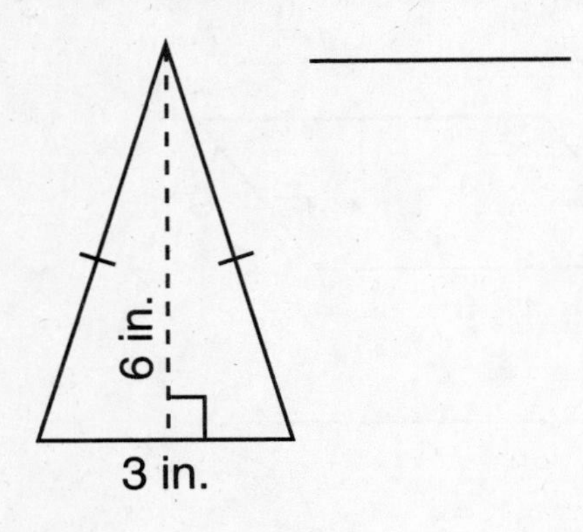

2. ___

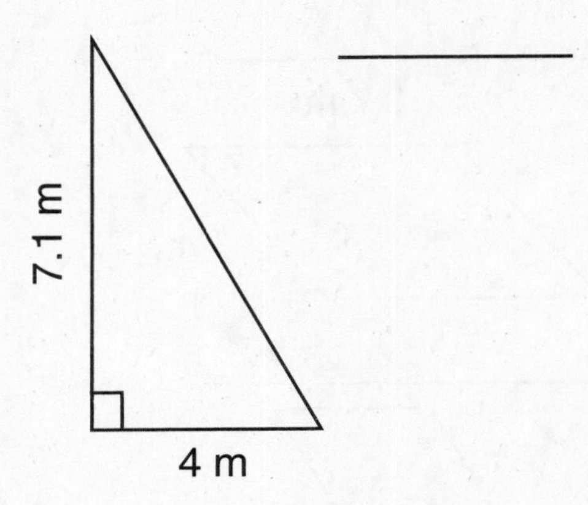

3. ___

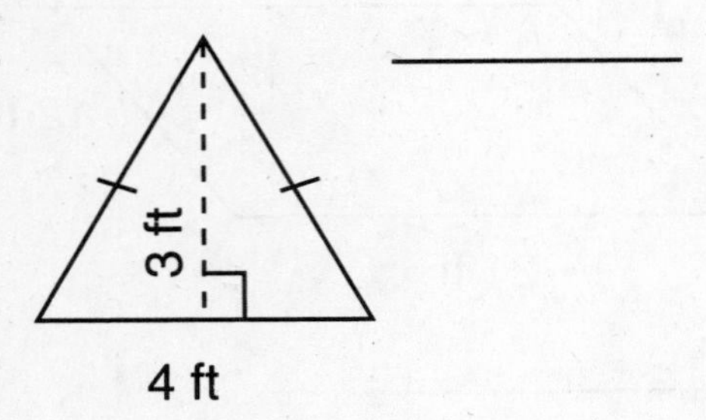

4. ___

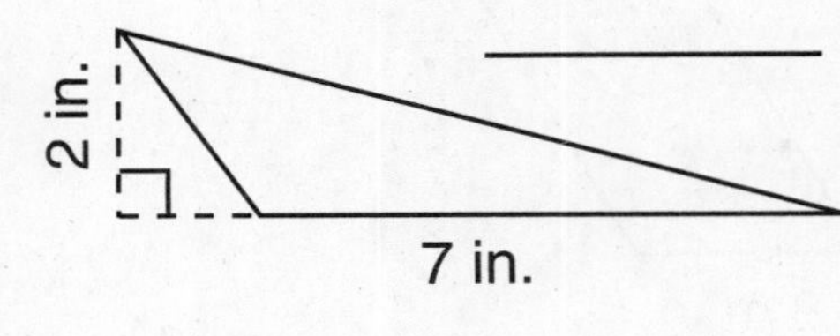

Solve.

5. At the museum, Julia sees a painting on a triangular canvas. The triangle's base is 4 feet, and its height is 2.5 feet. What is the area of the painting?

___

6. A park near Tyler's house has the shape of a right triangle with a base of 160 meters and a height of 200 meters. What is the area of the park?

___

Spiral Review

Tell whether the figures are congruent, similar, or neither.

7. ___

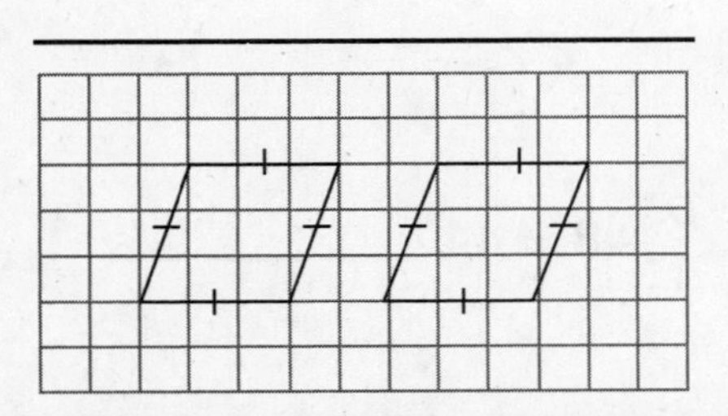

8. ___

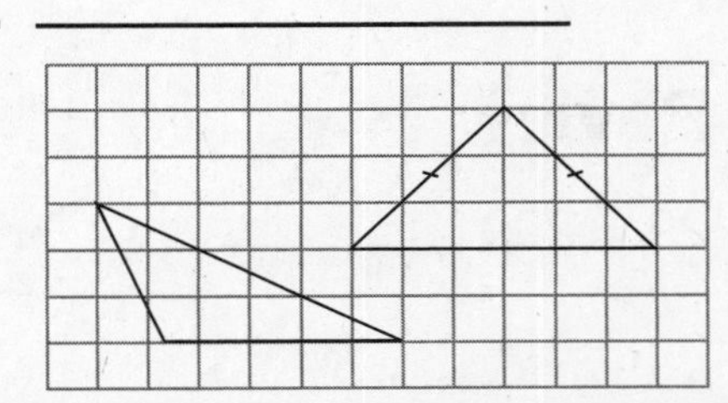

Name ____________________

Explore Circumference of Circles

Find the approximate circumference of each circle. Use × ≈ 3.14. Round to the nearest tenth, if necessary.

1. ____________________

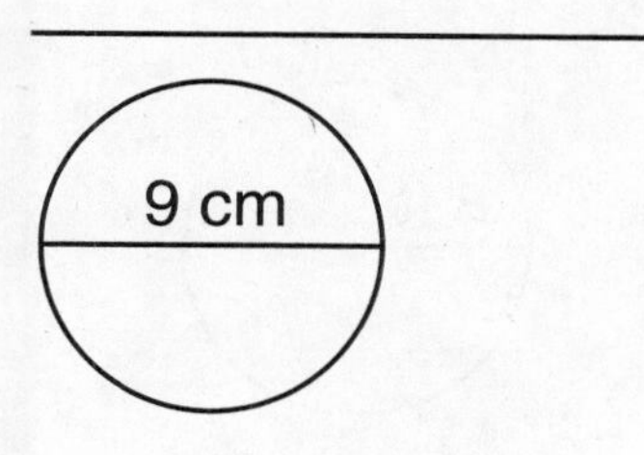

2. ____________________

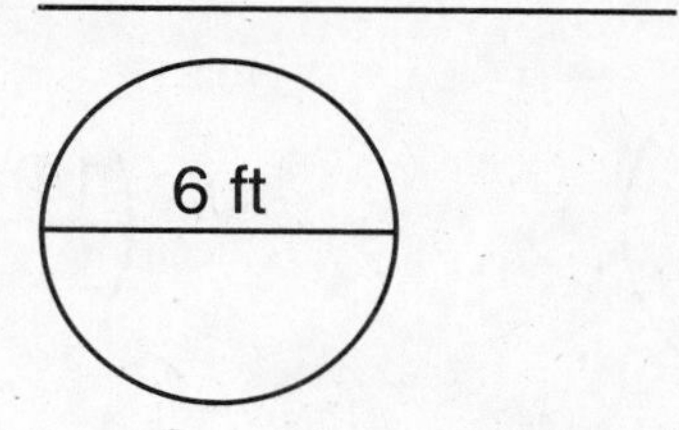

3. ____________________

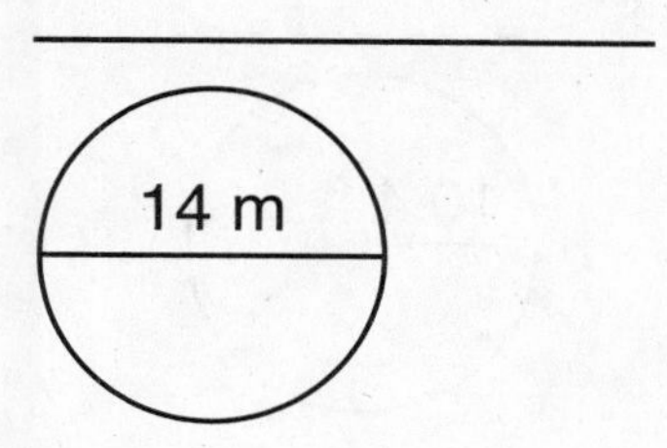

4. ____________________

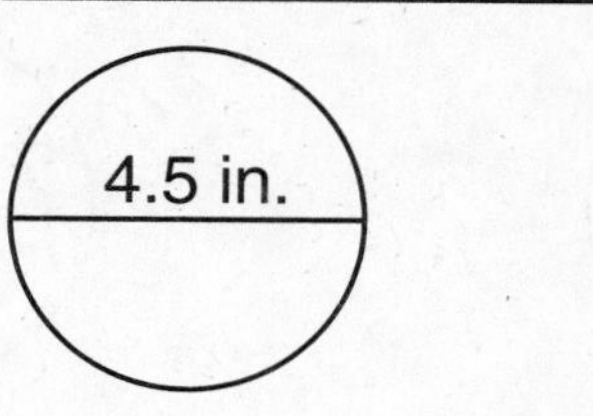

5. ____________________

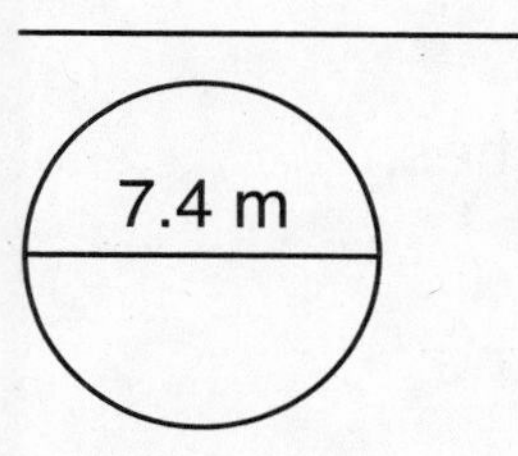

6. ____________________

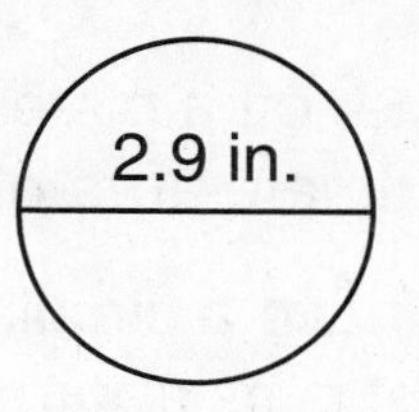

Solve.

7. Manuel and Fern order a large pizza from Cabot's Pizza. The pizza has a diameter of 16 inches. What is its approximate circumference? ____________________

8. The largest pie ever baked was a pecan pie 40 feet in diameter baked in Okmulgee, Oklahoma, in 1989. What was the pie's approximate circumference? ____________________

Spiral Review

Evaluate the expression for the value given.

9. $91 \times (h + 4)$ for $h = 5$ ________

10. $a + 625 \div 5$ for $a = 14$ ________

Name ____

Explore Area of Circles

Find the approximate area of each circle. Use $\pi \approx 3.14$. Round to the nearest tenth, if necessary.

1. ____________

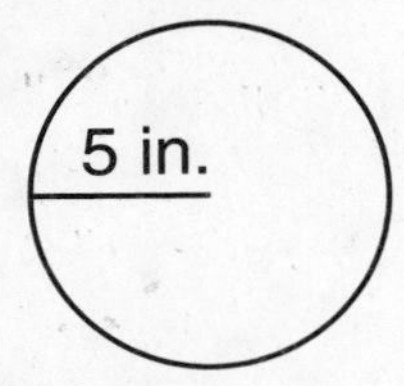

2. ____________

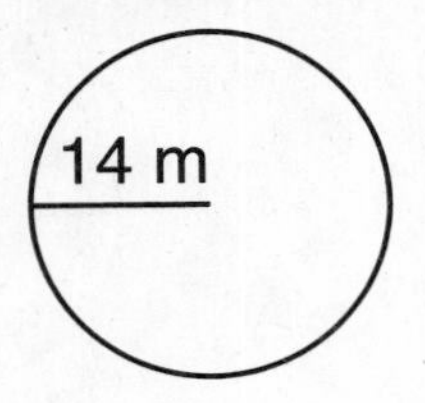

3. ____________

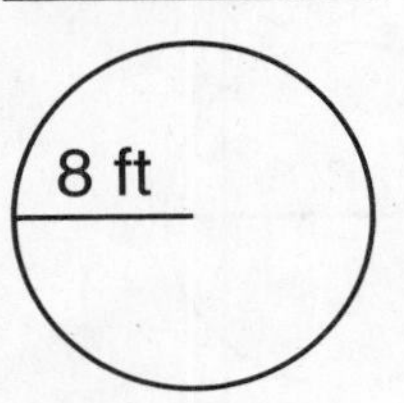

4. ____________

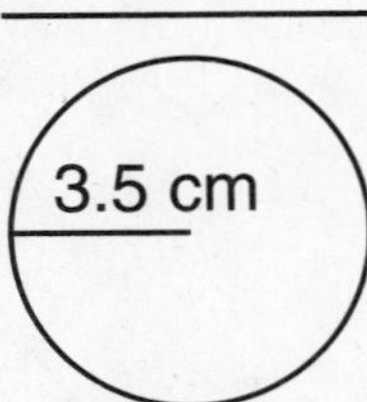

5. ____________

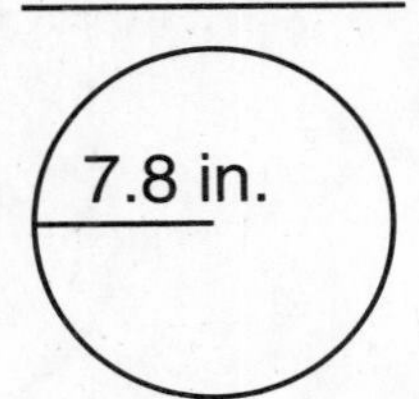

6. ____________

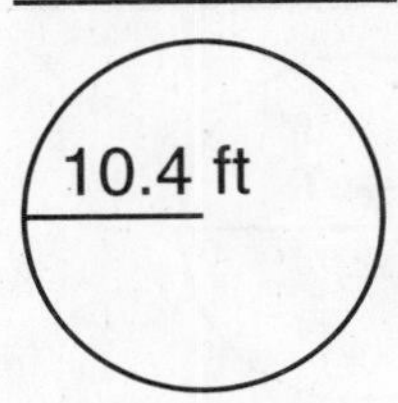

Solve.

7. Mary eats her dinner off a plate that has a radius of 4 inches. What is the approximate area of the plate? ____________

8. Roberto's parents have a circular table. The diameter of the table is 5 feet. What is its approximate area? ____________

Spiral Review

Write whether a translation, reflection, or rotation was made.

9. ____________________

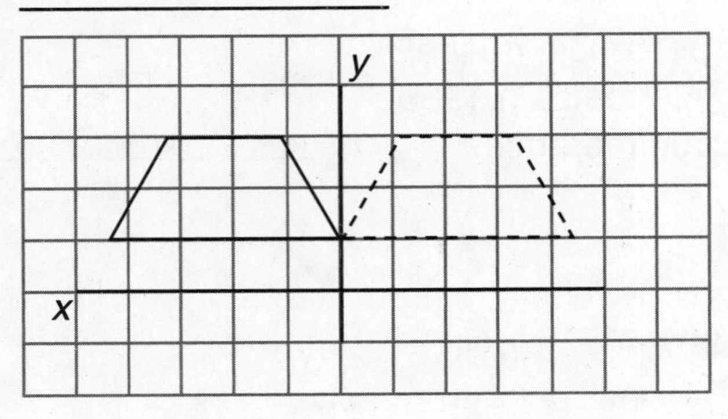

10. ____________________

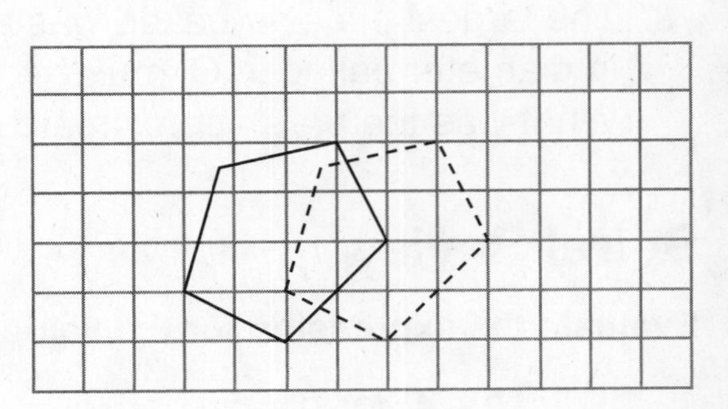

Name ______________________________

Problem Solving: Strategy
Solve a Simpler Problem

Solve. Explain how you simplified each problem.

1. In the middle of a square park there is a picnic shelter. The park is 80 meters on each side, and the picnic shelter has dimensions of 5 meters by 12 meters. What is the area of the park that is not covered by the picnic shelter?

____

____

2. The city decides to fence in the park completely except for an entrance that is 4 meters wide. How much fencing is needed?

____

3. The city buys a lot next to the park. The lot has the shape of a triangle with a base of 70 meters and a height of 40 meters. The city adds the lot to the park to create a larger park. What is the area of the larger park?

____

4. Tina is making a lowercase "i" out of colored paper. She makes a rectangle that measures 18 inches by 6 inches, and a circle with a radius of 3 inches. How much paper does she use in all?

____

5. Daniel's backyard measures 10 meters by 20 meters. He wants to build a circular patio in the backyard. The patio would have a radius of 3 meters. If he builds the patio, how much room will be left in the yard?

____

____

Spiral Review

6. $\frac{3}{7} = \frac{18}{\_\_\_}$
7. $\frac{4}{\_\_\_} = \frac{48}{108}$
8. $\frac{10}{11} = \frac{\_\_\_}{242}$
9. $\frac{\_\_\_}{15} = \frac{10}{75}$

Name ________________________________

3-Dimensional Figures and Nets

Write the number of faces, edges, and vertices for each figure.

1. ____________ **2.** ____________ **3.** ____________

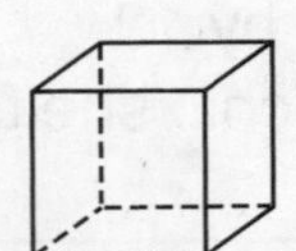

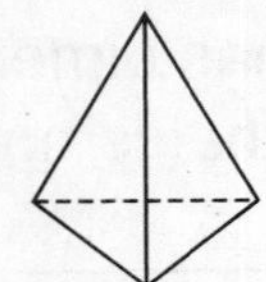

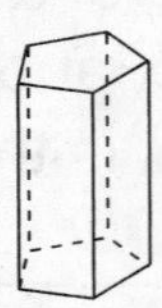

What 3-dimensional figure does each net make when cut and folded?

4. ____________ **5.** ____________ **6.** ____________

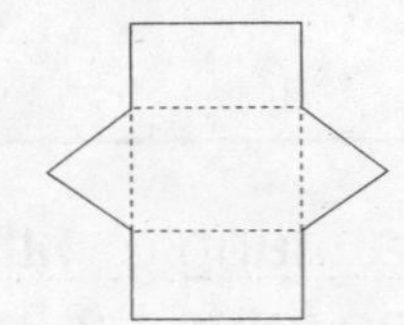

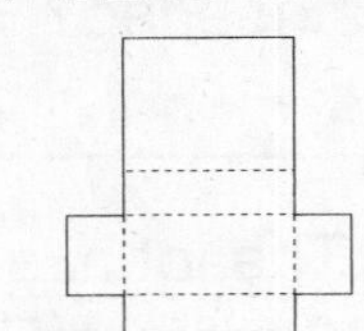

Draw the net for each 3-dimensional figure.

7. **8.** **9.**

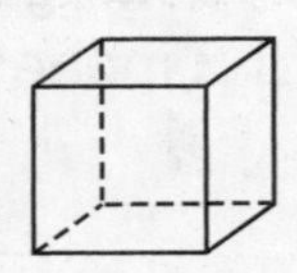

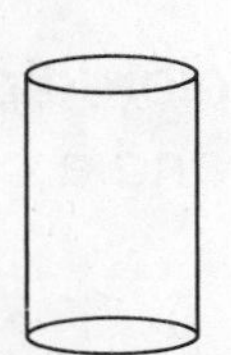

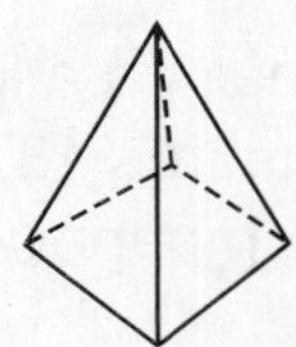

Problem Solving

10. Lou receives a package that is a cube with sides of 11 inches. What is the area of one of its faces?

____

11. Michael says that the net shown is for a sphere. Is he correct?

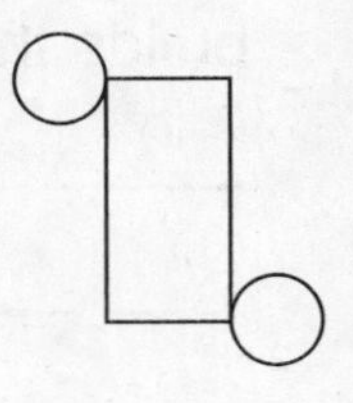

____

Spiral Review

Solve.

12. $2y + 5 = 431$ ____________

13. $\frac{u}{6} + 22 = 78$ ____________

14. $4n - 130 = 534$ ____________

15. $\frac{r}{25} + 45 = 56$ ____________

Name ________________________________

12·10 3-Dimensional Figures from Different Views

Draw the top view, the front view, and one side view of the shape.

1. 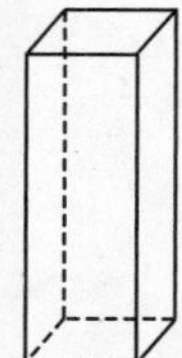

2. 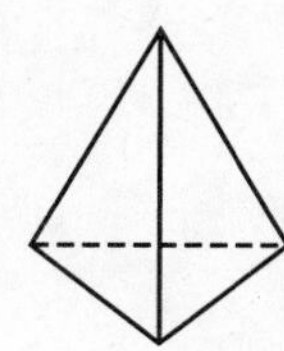

Problem Solving

3. Edna's Ice Cream uses cones like the one shown below. Is this view of the cone a side view or a top view? ____________

4. With one eye closed, Elise is looking at a 3-dimensional figure. All she can see of the figure is one square. Name at least one 3-dimensional figure that she could be looking at.

____

Spiral Review

Complete the table. Then graph the function.

5. $y = x - 2$

x	y
$^-2$	
0	
2	
4	

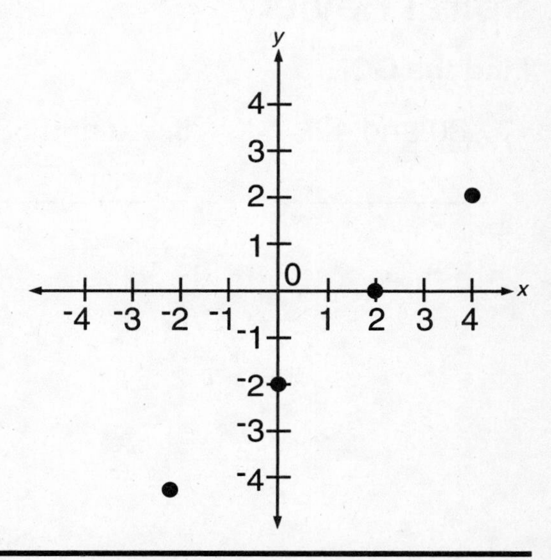

Name ______________________________

Explore Surface Area of Rectangular Prisms

Find the surface area of each rectangular prism.

1. ____________________

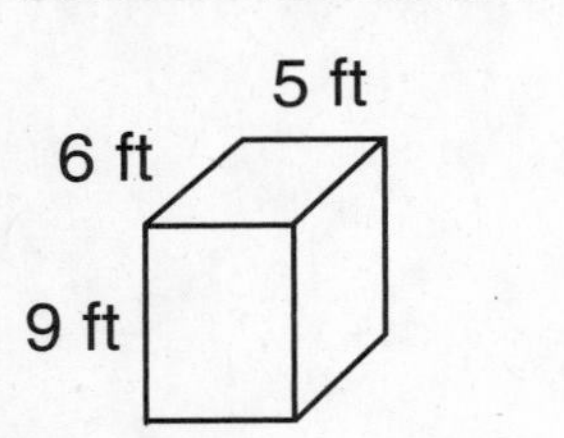

2. ____________________

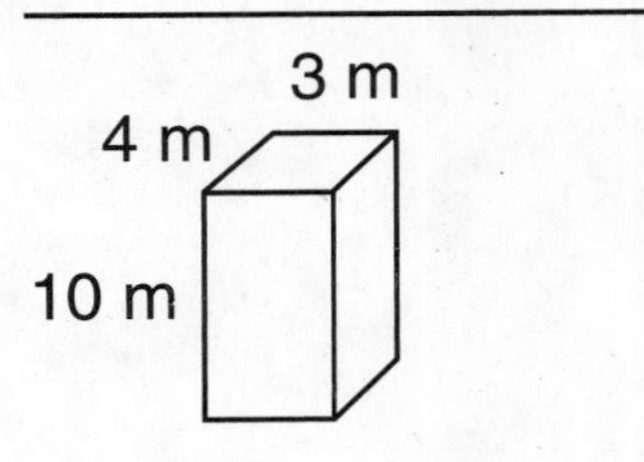

3. ____________________

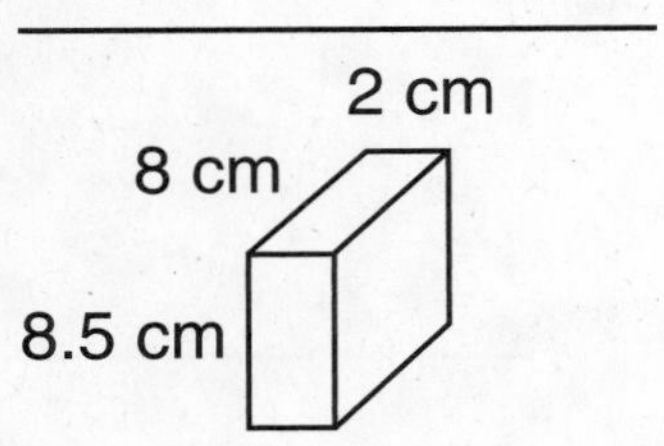

4. ____________________

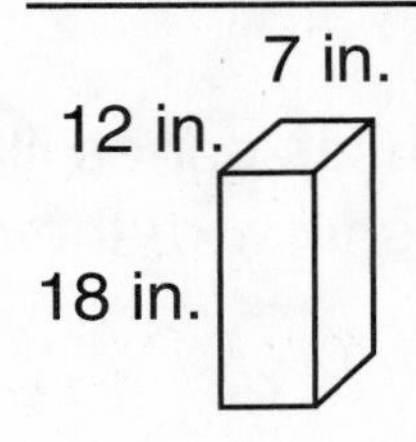

Solve.

5. Gram is wrapping a book that is 5 inches wide, 8 inches tall, and 2 inches thick. At least how much wrapping paper will he need to use to cover the entire book? ____________________

6. Erin's new refrigerator came in a box in the shape of a rectangular prism. The box is 6 feet tall, 2.5 feet wide, and 3 feet deep. What is the box's surface area? ____________________

Spiral Review

Find the GCF.

7. 30 and 42 ____________

8. 10 and 68 ____________

9. 24 and 132 ____________

10. 42 and 105 ____________

Name ______________________________

Explore Volume of Rectangular Prisms

Find the volume of each rectangular prism. Round to the nearest tenth, if necessary.

1. ______________

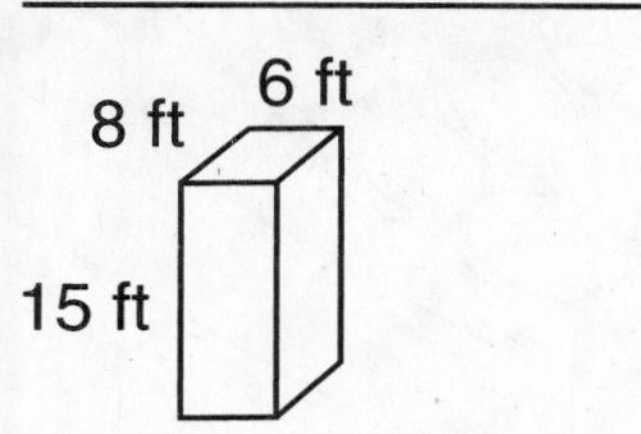

2. ______________

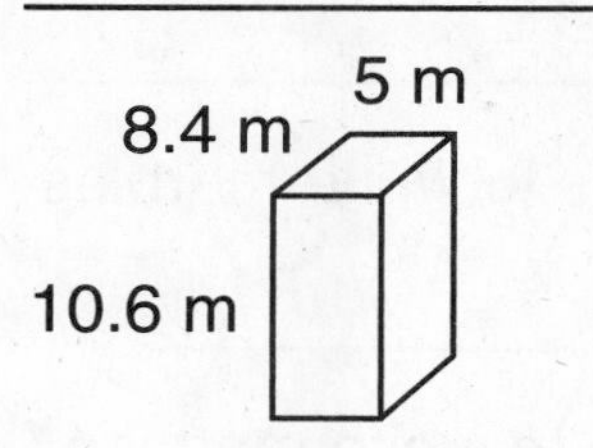

3. ______________

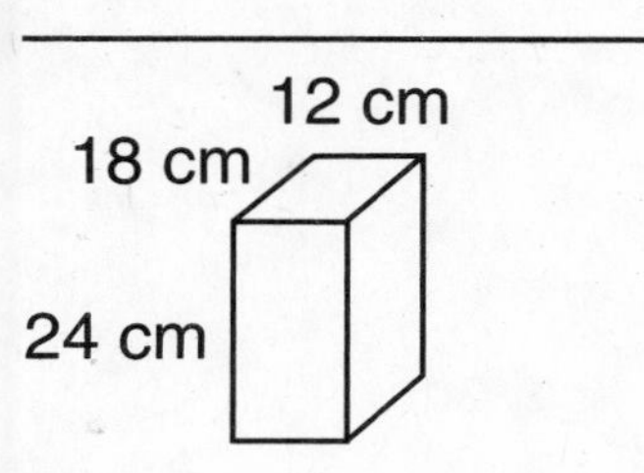

4. ______________

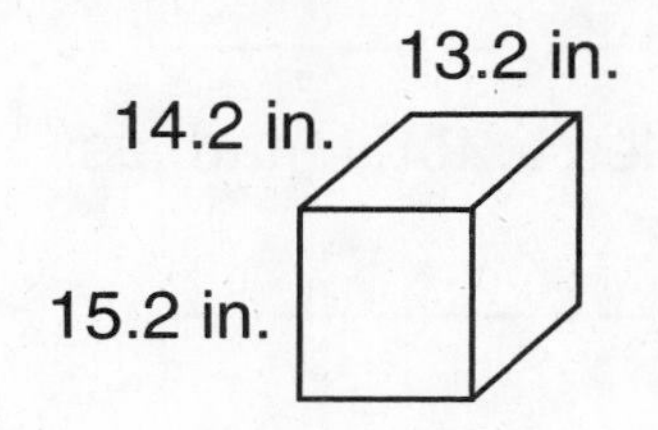

Solve.

5. Kyle is sending a package to his grandmother in a small box that is 8 cm wide, 6 cm tall, and 6 cm deep. What is the volume of the box? ______________

6. Teri has an aquarium that is 10 inches tall, 25 inches wide, and 12 inches deep. What is the aquarium's volume? ______________

Spiral Review

7. $\begin{array}{r} 241 \\ \times\ 1.6 \\ \hline \end{array}$

8. $\begin{array}{r} 560 \\ \times 7.2 \\ \hline \end{array}$

9. $\begin{array}{r} 1.56 \\ \times\ 4.4 \\ \hline \end{array}$

10. $\begin{array}{r} 314 \\ \times 822 \\ \hline \end{array}$

Name ___________________________________

Explore Ratio

Write each ratio in three ways.

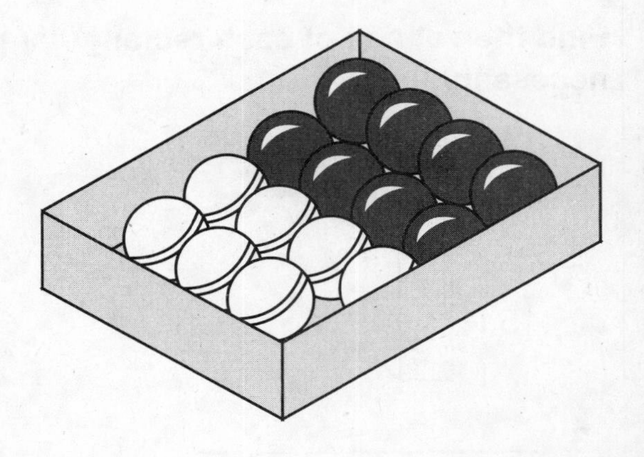

1. white marbles to black marbles

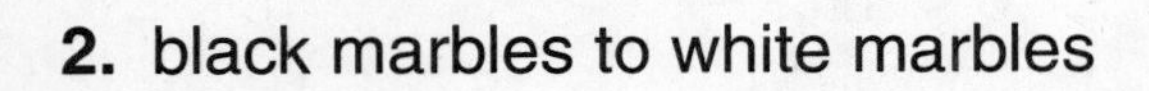

2. black marbles to white marbles

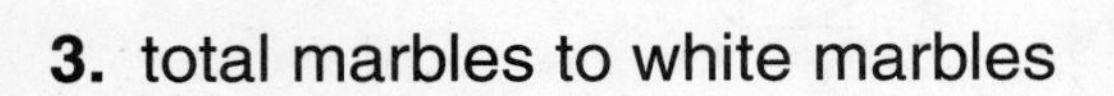

3. total marbles to white marbles

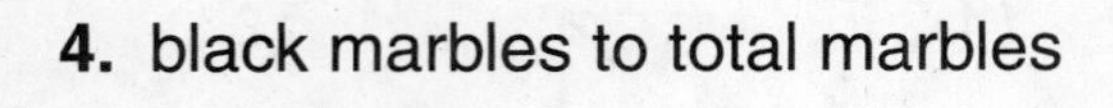

4. black marbles to total marbles

____

Solve.

5. Maia notices that there are 3 hot dogs and 5 hamburgers on the grill. What is the ratio of hamburgers to hot dogs?

____

6. Greg buys 4 peach yogurts and 6 plain yogurts. Is the ratio 10 to 4 the ratio of total yogurts to peach yogurts, peach yogurts to plain yogurts, or peach yogurts to total yogurts?

____

Spiral Review

Find the area.

7.

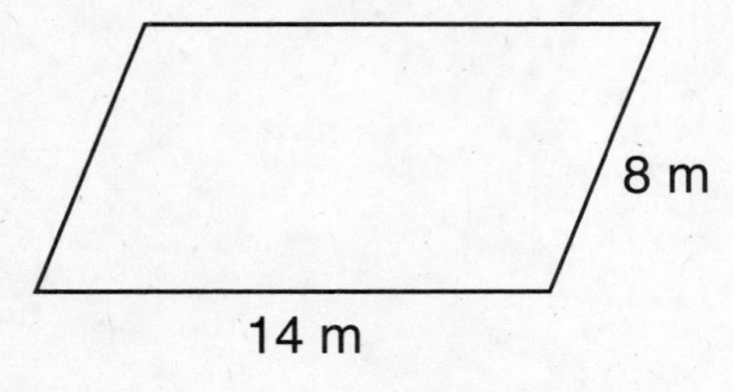

____

8.

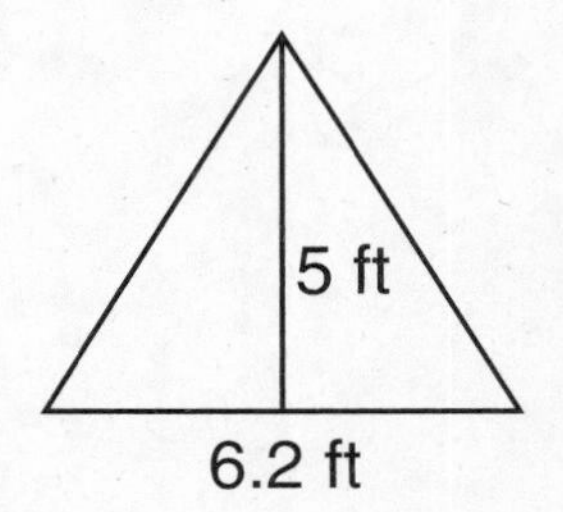

____

Name ____________________

13-2 Equivalent Ratios

Complete each ratio table.

1.

4	8			20
9		27	36	

2.

12			48	60
7	14	21		

Tell whether the ratios are equivalent. Write *Yes* or *No*.

3. $\frac{4}{5}, \frac{24}{30}$ ________

4. $\frac{7}{4}, \frac{42}{24}$ ________

5. $\frac{27}{3}, \frac{9}{2}$ ________

6. $\frac{5}{11}, \frac{25}{55}$ ________

7. 9:10, 19:20 ________

8. 36:81, 4:9 ________

9. 3:8, 36:96 ________

10. 24:12, 8:3 ________

Name three ratios equivalent to the given ratio.

11. $\frac{3}{4}$ ____________

12. $\frac{7}{2}$ ____________

13. $\frac{25}{30}$ ____________

Find the missing number.

14. 3:5 = x:25 ________

15. 8:1 = u:9 ________

16. 35 to 21 = r to 3 ________

Problem Solving

17. The ratio of girls to boys in Ms. Needham's class is 4:5. Find an equivalent ratio. ____________

18. There are two fruit baskets for sale at the grocery store. One has 3 bananas and 7 kiwi fruit. The other has 9 bananas and 14 kiwi fruit. Are the ratios of bananas to kiwi fruit in the baskets equivalent? ____________

Spiral Review

19. $1\frac{1}{4} \times 1\frac{1}{5} =$ ______

20. $6\frac{2}{3} \div \frac{5}{6} =$ ______

21. $2\frac{1}{9} \div 1\frac{1}{3} =$ ______

Name ____________________

Rates

Complete.

1. 175 km in 2 h = _____ km in 8 h
2. 16 oz in 3 min = _____ oz in 15 min
3. 13 people at 2 tables = _____ people at 16 tables
4. $81.35 in 8 hours = _______ in 40 hours

Find each unit rate.

5. $4.50 for 5 lb = _____ for 1 lb
6. 640 pages in 8 notebooks = _____ pages in 1 notebook
7. 504 tissues in 9 boxes = _____ tissues in 1 box
8. 384 mi on 12 gal = _____ mi on 1 gal

Decide which is the better buy.

9. 4 shirts for $24
   5 shirts for $35
   ____
10. 7 CDs for $77
    11 CDs for $99
    ____
11. 9 bicycles for $2,700
    6 bicycles for $2,000
    ____

Problem Solving.

12. It costs $6.40 to add 4 toppings to a large pizza from Rick's Pizza House. What is the rate per topping? ________
13. During a storm, 1.5 inches of rain falls in one minute. If rain continues to fall at this rate for 10 minutes, how much rain will fall? ________

Spiral Review

Find the approximate circumference. Round to the nearest tenth. Use $\pi = 3.14$.

14. ________

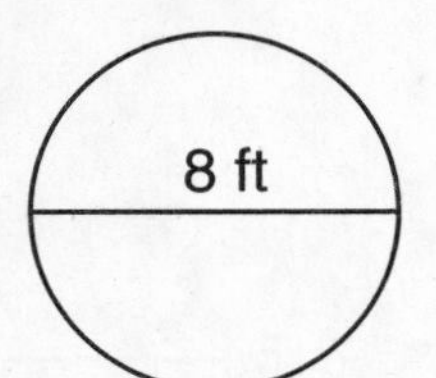

15. ________

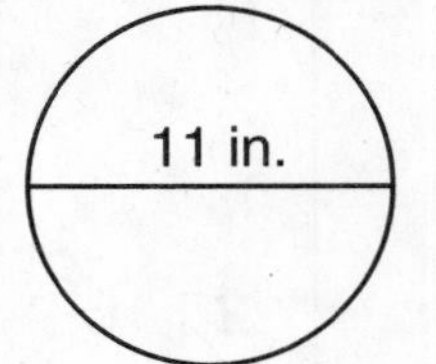

Name ____________________

Problem Solving: Reading for Math

Check the Reasonableness of an Answer

Use data from the table for problems 1–4. Explain your answer.

1. The table shows how long it takes Marisa and David to do different chores. On Saturday, Marisa is supposed to clean both bathrooms in her family's house. She calculates that it will take her about an hour and a half. Is her calculation reasonable?

Chore	Time
Taking out Trash	3 min
Cleaning Bathroom	45 min
Sorting Recycling	20 min
Vacuuming	1 h
Unloading Dishwasher	10 min
Weeding	2 h

____

____

2. On the same Saturday, David is supposed to take out the trash, vacuum, and unload the dishwasher. He calculates that he has to work longer than Marisa. Is his calculation reasonable?

____

3. Marisa weeds the garden twice a month during June, July, and August. She calculates that in all, she spends about half a day of her summer vacation weeding. Is her calculation reasonable?

____

____

4. David decides to start a neighborhood recycling service. He plans to charge each neighbor $3.00 for sorting his or her recycling once a week. He calculates that if each neighbor's recycling takes him as long as his family's recycling, he will make about $9.00 an hour. Is his calculation reasonable?

____

____

Spiral Review

5. 312×4.5

6. 7.54×80

7. 1.65×6.4

8. 10.933×28

Name ____________________

Scale Drawings

Use data from the map for problems 1–3. Find each actual size.

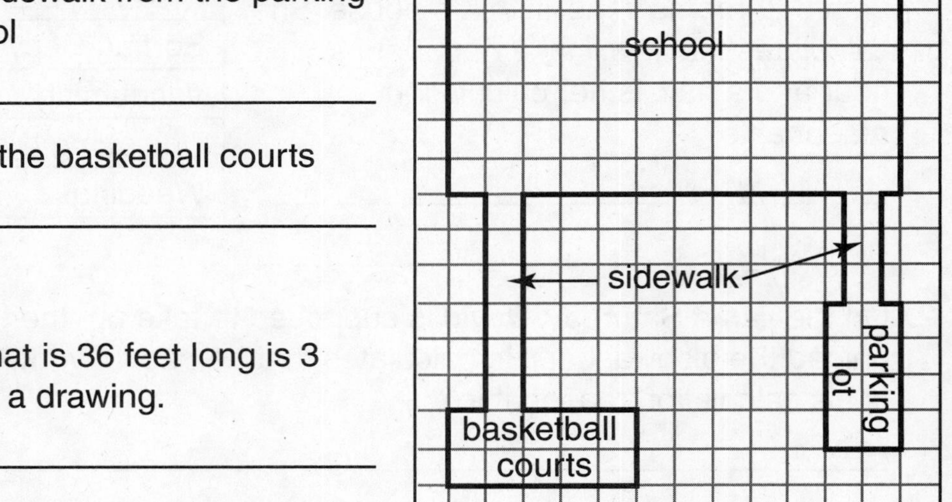

1. perimeter of the school building ____________________

2. length of the sidewalk from the parking lot to the school ____________________

3. dimensions of the basketball courts ____________________

Find the scale.

4. A classroom that is 36 feet long is 3 inches long on a drawing. ____________________

5. A playground that is 45 meters wide is 5 cm wide on a drawing. ____________________

Problem Solving

6. Debbie is looking at a map of Minnesota with a scale of 1 inch = 50 miles. She measures 3 inches on the map from her home town to the border between the United States and Canada. What is the actual distance? ____________________

7. A wall map shows an area of 4 kilometers by 6 kilometers in a space 2 meters by 3 meters. What is the map's scale? ____________________

Spiral Review

Solve.

8. $p - 314 = 315$ ________ **9.** $\frac{x}{3} - 12 = 10$ ________ **10.** $4c + 14 = 258$ ________

Name ______________________________

Explore Probability

Use the spinner for problems 1–4.

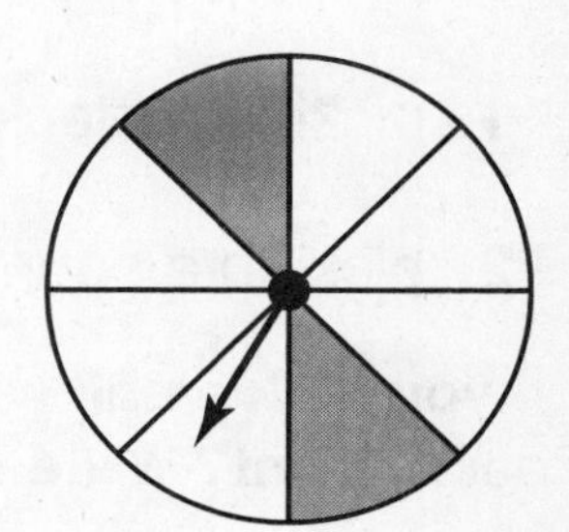

1. What are the possible outcomes?

   ____

2. Which outcome is unlikely? ______________________

   Which outcome is likely? ______________________

3. If you spin the spinner 60 times, which outcome do you think you will get more often? ______________________

4. Suppose you change 4 of the white sections to gray. Now which outcome is unlikely? ______________________

Solve.

5. Caleb has a spinner with 9 orange sections and 1 blue section. If he changes 1 of the orange sections to blue, which outcome will be likely? ______________________

6. Laura makes a spinner with 12 sections. One of the sections is red, a second is yellow, and a third is green. The rest are blue. When she spins, is it likely or unlikely that the outcome will be a color other than blue? ______________________

Spiral Review

Find the approximate area of each circle. Round to the nearest tenth if necessary. Use $\pi = 3.14$.

7.

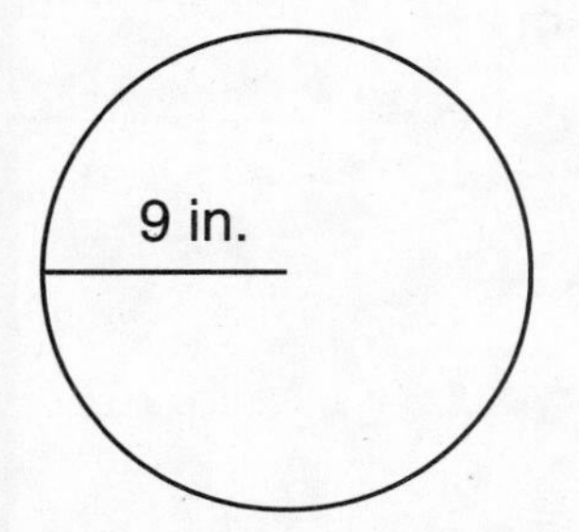

____

8.

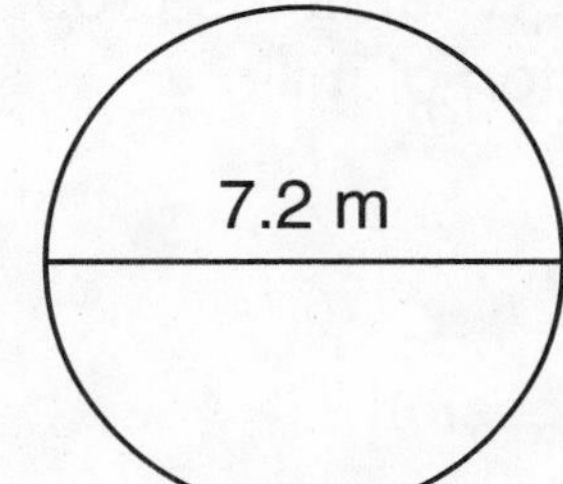

____

Name ______________________________

Probability

If you pick a counter without looking, what is the probability of each event?

1. picking a black counter ________

2. picking a counter with an "X" ______________

If you pick a ball without looking, what is the probability of each event? Write *certain* or *impossible* for each event.

3. picking a volleyball

   ____

4. picking a ball

   ____

If you spin the spinner, what is the probability of each event? Write *more likely than, less likely than,* or *equally likely* to complete each sentence.

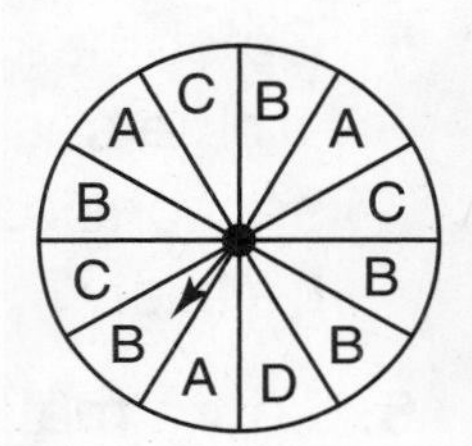

5. Spinning a D is ______________________________ spinning a B.

6. Spinning a C is ______________________________ spinning an A.

Problem Solving

7. In a cooler are 4 bottles of juice, 3 bottles of water, and 5 bottles of lemonade. If Christine picks without looking, what is the probability that she will pick lemonade? ______________

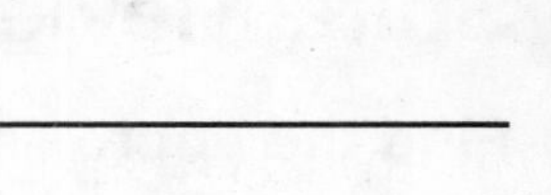

8. Before Christine picks, someone puts 3 cartons of milk into the cooler. What is the probability that Christine will now pick a carton of milk? ______________

Spiral Review

Simplify each expression.

9. $4 \times 13 + 12 \div 3 =$ ________

10. $126 + 14 \div 7 =$ ________

11. $6 \times 5 \times 2^2 =$ ________

11. $140 - (3^2 + 3 \times 7) =$ ________

Name ______________________________

Problem Solving: Strategy

Do an Experiment

Do an experiment to solve. Record the results in the frequency table below.

1. A dollar bill is dropped face-up from a height of 3 feet. What is the probability that it will land face down? First make and record a prediction. Then record the results of dropping a dollar bill 50 times.

Prediction: ______________________

How It Lands	Tally	Frequency
Faceup		
Facedown		

2. Which punctuation mark is the most commonly used? First make and record a prediction. Then choose 40 sentences from a book and record the occurrence of punctuation.

Prediction: ______________________

Punctuation Mark	Tally	Frequency
Period		
Question Mark		
Exclamation Point		
Comma		

Mixed Strategy Review

3. Angela wants to order a tool set that costs \$44.95 plus \$2.25 for shipping. She has 160 quarters in her bank. Does she have enough money? Explain.

____

4. Ralph has a $6\frac{1}{4}$ -ounce jar of metallic paint. He uses $1\frac{3}{4}$ ounces to paint one model and $2\frac{1}{2}$ ounces to paint another. How much paint is left? ______

Spiral Review

5. $\frac{7}{2} \times 6 =$ ______

6. $\frac{3}{4} + \frac{5}{3} =$ ______

7. $1.5 \times 4.2 =$ ______

8. $3\frac{3}{4} \times 7\frac{1}{3} =$ ______

9. $\frac{8}{9} \div \frac{12}{5} =$ ______

10. $3\frac{3}{5} - \frac{2}{3} =$ ______

Name ____________________

Explore the Meaning of Percent

Write each fraction, decimal, or ratio as a percent.

1. $\frac{1}{4}$ ______ **2.** 0.40 ______ **3.** 50:100 ______

4. $\frac{36}{100}$ ______ **5.** $\frac{6}{10}$ ______ **6.** 0.73 ______

7. $\frac{2}{5}$ ______ **8.** 0.14 ______ **9.** $\frac{9}{10}$ ______

Solve.

10. A 10 × 10 grid has 54 squares shaded blue. What is the percent of squares that are shaded blue? ______

11. In a survey, 5 out of 10 residents of Silo City say they are happy with the job that their mayor is doing. What is the percent of residents surveyed who are happy with the mayor? ______

12. In a basket of 5 apples, 3 are red and 2 are yellow. What is the percent of apples that are red? ______

Spiral Review

Find the area.

13.

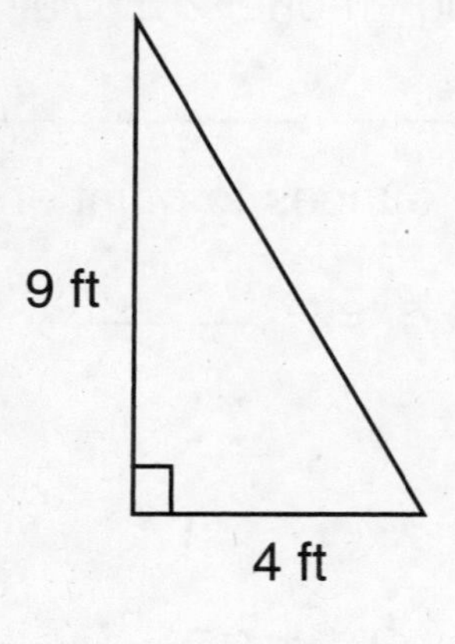

______

14.

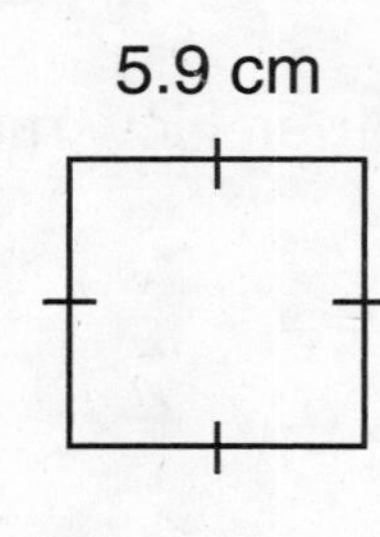

______

15.

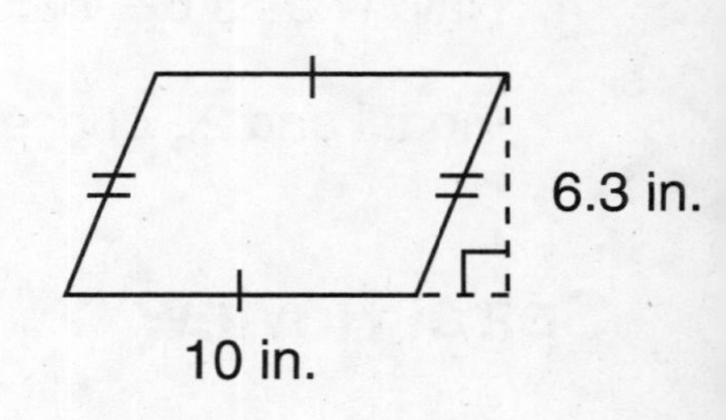

______

Name ___________________________________

Percents, Fractions, and Decimals

Write each percent as a decimal and as a fraction in simplest form.

1. 20% ________ **2.** 70% ________ **3.** 35% ________

4. 55% ________ **5.** 25% ________ **6.** 30% ________

7. 90% ________ **8.** 47% ________ **9.** 63 % ________

Write each fraction or decimal as a percent.

10. $\frac{2}{5}$ ________ **11.** 0.24 ________ **12.** $\frac{6}{10}$ ________

13. 0.76 ________ **14.** $\frac{1}{2}$ ________ **15.** 0.51 ________

16. $\frac{3}{4}$ ________ **17.** $\frac{81}{100}$ ________ **18.** 0.93 ________

Problem Solving

19. Colin takes 10 of his paintings to an art show and sells 2 of them. How can you write this number as a percent? ____________

20. Greta takes 20 shots during a basketball game and makes 11 of them. How can you write this number as a percent? ____________

Spiral Review

Find each unit rate.

21. \$3.60 for 5 lb = ________ for 1 lb

22. 264 seats in 8 rows = ________ seats in 1 row

23. \$34.95 for 5 book = ________ for 1 book

24. 110 mi in 2 hours = ________ in 1 hour

25. 176 mi on 8 gal = ________ mi on 1 gal

Name ______________________________

More About Percents

Write each percent as a decimal and as a mixed number in simplest form, or as a whole number.

1. 125% ________ **2.** 300% ________ **3.** 240% ________

4. 500% ________ **5.** 370% ________ **6.** 425% ________

7. 950% ________ **8.** 800% ________ **9.** 680% ________

Write each percent as a decimal and as a fraction in simplest form.

10. 32% ________ **11.** 60% ________ **12.** 36% ________

13. 21% ________ **14.** 56% ________ **15.** 84% ________

16. 37.5% ________ **17.** 74% ________ **18.** 15% ________

Problem Solving

19. Of the 8 houses on Danny's street, only 1 is painted blue, for a ratio of 1:8. Write this ratio as a percent and as a fraction in simplest form. ________

20. Over the last 20 years, the population of Cactusville has grown by 225%. Write this percent as a decimal and as a mixed number in simplest form. ________

Spiral Review

Solve.

21. $\frac{q}{8} + 13 = 120$ ________ **22.** $2r + 4 = {}^{-}10$ ________

23. $3s - 50 = {}^{-}5$ ________ **24.** $\frac{t}{10} + 8 = 11$ ________

Name ______________________________

Problem Solving: Reading for Math

Represent Numbers

Solve.

1. A survey asks customers of the Super Grocery Center what their favorite fruits are. Of the people surveyed, $\frac{1}{4}$ say apples, 28% say bananas, 0.29 say oranges, and 18% say mangoes. How would you list these fruits from least to greatest number of responses?

____

2. Did more than $\frac{1}{8}$ of those surveyed say mangoes are their favorite fruit?

____

3. Charles is working to improve his free-throw shooting in basketball. In the first game of the season, he shoots 40% from the free-throw line. By the last game of the season, he makes 8 of the 10 free throws he attempts. Does he improve his shooting?

____

Use data from the table for problems 4–7.

1860 Presidential Election Results

Candidate	Percent of Popular Vote
Stephen Douglas	29%
John Breckinridge	18%
Abraham Lincoln	40%
John Bell	13%

4. Of the four candidates listed, which one received the most popular votes?

____

5. Who received more popular votes, John Breckinridge or John Bell?

____

6. List the candidates from most popular votes to least.

____

7. Write Lincoln's percent of the popular vote as a decimal and as a fraction in its simplest form. ______________

Spiral Review

8. 6.27 – 0.88 = ________ **9.** 1.45 + 17.2 = ________ **10.** 4.68 × 3.1 = ________

Name ______________________________

Percent of a Number

Find the percent of each number.

1. 40% of 60 ________ **2.** 50% of 90 ________

3. 10% of 50 ________ **4.** 300% of 40 ________

5. 20% of 20 ________ **6.** 80% of 70 ________

7. 25% of 64 ________ **8.** 150% of 50 ________

9. 210% of 80 ________ **10.** 35% of 60 ________

11. 75% of 120 ________ **12.** 40% of 70 ________

13. 40% of $7.00 ________ **14.** 15% of $4.00 ________

15. 35% of $6.00 ________ **16.** 250% of $3.00 ________

17. 20% of $6.00 ________ **18.** 200% of $0.40 ________

19. 80% of $1.00 ________ **20.** 110% of $0.70 ________

Problem Solving

21. Of the 60 students who sign up for basketball, 65% say that they have played the game before. How many students say that they have played basketball before? ________

22. Of the 28 students who sign up for softball, 25% want to be pitchers. How many want to be pitchers? ________

Spiral Review

23. 43 students in 2 buses = ________ students in 8 buses

24. 15 mi in 12 min = ________ mi in 1 h

25. 6 pies for $21.60 = ________ pies for $54.00

26. 32 oz in 2 qt = ________ oz in 4 gal

27. 17 rabbits in 3 cages = ________ rabbits in 12 cages

28. 5 doz egg whites in 6 cakes = ________ egg whites in 8 cakes

Name ____________________

Percent That One Number Is of Another

Find the percent each number is of the other. Round to the nearest whole number percent, if necessary.

1. What percent of 80 is 20? ________

2. What percent of 30 is 9? ________

3. 18 is what percent of 90? ________

4. What percent of 50 is 35? ________

5. What percent of 70 is 56? ________

6. 81 is what percent of 90? ________

7. What percent of 40 is 22? ________

8. 90 is what percent of 60? ________

9. What percent of 120 is 30? ________

10. What percent of 80 is 160? ________

Problem Solving

11. Asa goes to the store to buy 5 grapefruit. He winds up buying a bag of 6 grapefruit. What percent of 5 is 6? ________

12. Celeste needs to collect 40 different kinds of leaves for a class project. So far, she has collected 32 different kinds. What percent of her total goal has she collected so far? ________

Spiral Review

Classify each triangle as equilateral, isosceles, or scalene and as right, acute, or obtuse.

13.

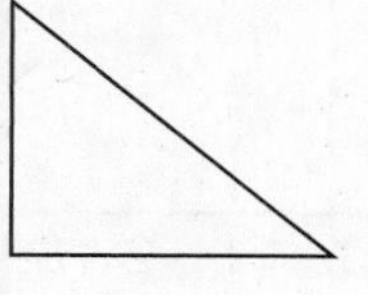

____

14.

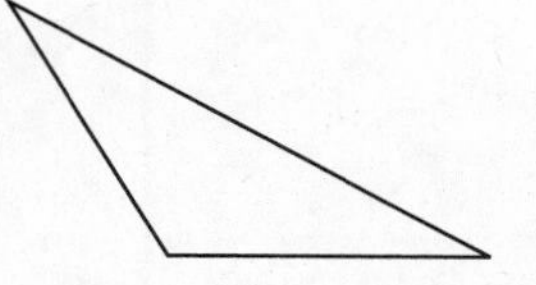

____

15.

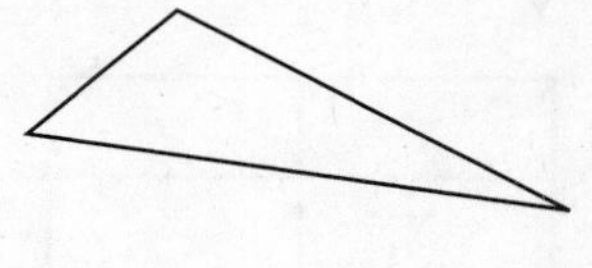

____

Name ______________________________

Problem Solving: Strategy
Use Logical Reasoning

Use a Venn diagram to solve each problem.

1. Of 28 students in Jean's class, 19 have signed up for volleyball and 12 have signed up for basketball. Three have signed up for both sports. How many students have signed up for volleyball but not basketball?

   ____

2. Of 15 children in a neighborhood, 10 like to ride bicycles and 8 like to rollerblade. Three children like to do both. How many children like to rollerblade but don't like to ride bicycles?

   ____

3. Of 55 students in the fifth grade, 43 are fans of the local basketball team and 35 are fans of the local hockey team. Twenty-three students are fans of both teams. How many students are fans of the hockey team but not the basketball team?

   ____

4. Of 30 runners surveyed, 22 go running in the morning and 15 go running in the evening. Seven of the runners surveyed go running in both the morning and the evening. How many runners go running in the morning but not the evening?

   ____

Spiral Review

Complete the table. Then graph the function.

5.

$y = x + 3$

x	y
$^-4$	
$^-2$	
0	
2	

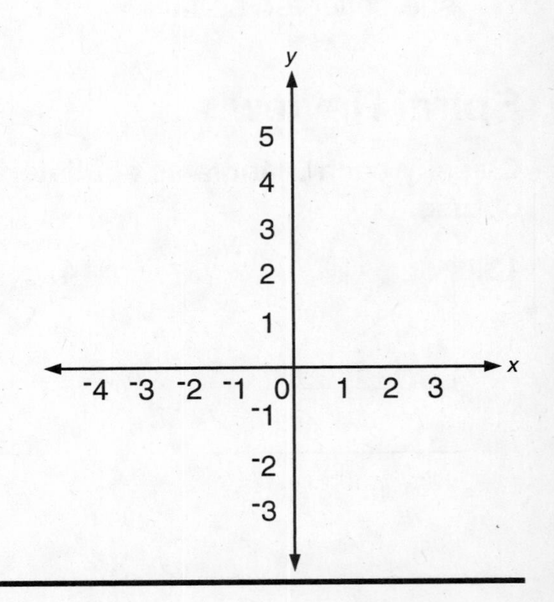

Name______________________________________

Circle Graphs

Use data from the circle graph for problems 1–4.

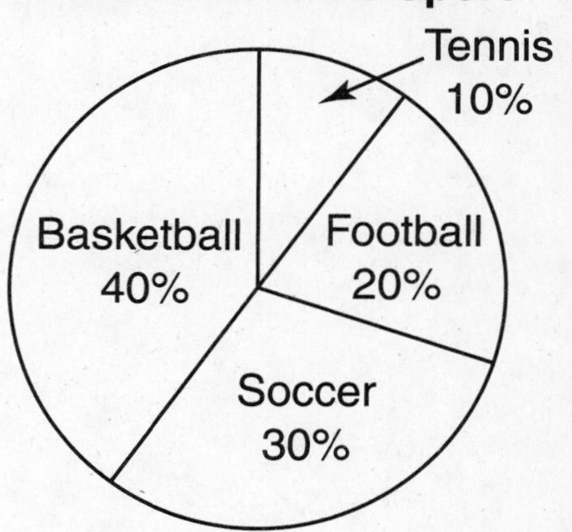

1. List the sports from least favorite to favorite.

   ____

2. What fraction of the total votes did football receive?

   ____

3. If 150 people were surveyed, how many voted for basketball?______________

Problem Solving

Use data from the table for problems 4–6.

Favorite Beach Activity	Percent of Responses
Walking	20%
Swimming	50%
Volleyball	10%
Reading	20%

4. List the number of degrees that each answer will occupy on a circle graph.

   ____

   ____

5. Make a circle graph to show the data.

6. If 320 people were surveyed, how many people said swimming was their favorite beach activity? ______________

Spiral Review

Tell whether the ratios are equivalent.

7. 4:5, 48:60 ________ 8. 120:45, 8:3 ________ 9. 6:7, 25:28 ________